Summer Solutions.
Minutes a Day-Mastery for a Lifetime!

Level 7

Problem Solving

Donna Marie Mazzola

Bright Ideas Press, LLC
Cleveland, OH

Summer Solutions Level 7
Problem Solving

All rights reserved. No part of this publication may be reproduced or transmitted in any form or by any means, electronic or mechanical, including photocopy, recording, or any information storage or retrieval system. Reproduction of these materials for an entire class, school, or district is prohibited.

Printed in the United States of America

ISBN-13: 978-1-934210-57-4
ISBN-10: 1-934210-57-9

Cover Design: Dan Mazzola
Editor: Kimberly A. Dambrogio
Illustrator: Christopher Backs

Copyright © 2010 by Bright Ideas Press, LLC
Cleveland, Ohio

Summer Solutions Level 7
Problem Solving

Lesson	Topic	Page
1	Strategy: Guess and Check	4
3	Strategy: Work Backward	8
5	Strategy: Look for a Pattern	12
7	Strategy: Use Logical Reasoning	16
9	Strategy: Make an Organized List	20
11	Strategy: Solve a Simpler Problem	24
13	Strategy: Use a Table / Make a Table	28
15	Strategy: Write an Equation	32
17	Strategy: Make a Model	36
19	Strategy: Draw a Picture or Diagram	40
	Help Pages	65
	Answers to Lessons	87

Dear Students,

I am Math-U and this is my friend, Add-E. We'll be around every summer to help you with Problem Solving. Math problems with words can be complicated sometimes, and having a collection of strategies to choose from will boost your confidence with problem solving. You practiced lots of strategies during the past year, and we don't want you to forget them. So, this summer we'll review ten strategies. Look for suggestions and hints on each page and even in your work boxes. Also, notice the thinking processes I'm modeling. I'll be there to help you be the best math problem solver you can be. Have a great summer!

P.S. It's best to do 3 lessons every week of your summer. Pick 3 days, like Monday, Wednesday, and Friday, to practice these skills. That way you won't feel foggy in the fall when school starts. It'll be fun!

Lesson #1

Strategy: **Guess and Check**

The best strategy to use when solving a jigsaw puzzle is to check different pieces until the one that fits is found. For the **guess and check** strategy, make a guess. See if it fits all the clues. If it does, the item is solved. If it does not, let that idea improve the next guess.

Example:

The length of a rectangle is 8 inches longer than its width. The area of the rectangle is 33 square inches. What is the length of the rectangle? What is the width?

Find what you know. Plan what to do. Ask if the answer is reasonable.

- The length is 8 inches more than the width.
- To find the area of a rectangle, multiply the length by the width.
- Guess and check numbers that have a difference of 8.

Length	Width	Difference	Area
9	1	8	9
10	2	8	20
11	3	8	33

- The last set of numbers fits all the clues: a difference of 8 and a product of 33. This answer is reasonable.

Final answer: The length of the rectangle is 11 inches and the width is 3 inches. These two numbers have a difference of eight. A rectangle with these dimensions would have an area of 33 square inches.

Use the **guess and check** strategy for the next four problems.

Summer Solutions© Problem Solving Level 7

Use these boxes to show your work.

1. The length of a rectangle is 24 inches longer than its width. The area of the rectangle is 145 square inches.

    What is the length of the rectangle? What is the width?

    1.

2. To find an average, add all the values, and then divide by the number of addends.

    The average, or mean, of which two values below is 222?

    237   111   333   407

    2.

3. I'm thinking of two numbers whose sum is 29 and whose product is 100.

    What are the two numbers?

    3.

4. The base of a parallelogram is 5 cm longer than its height. The area of the parallelogram is 300 cm².

    What is the measure of the base of the parallelogram? What is the height?

    4.

Lesson #2

1 – 2. An **outlier** is a number in a set of data that is either much larger or smaller than the other numbers in the group. An outlier can move the mean away from a group of numbers that has a small range.

Name	Spending Money
Jill	$14
Kaye	$15
Val	$50
Peg	$16
Sue	$15

What is the average amount of spending money these girls brought on a school field trip?

_____

Which amount is the outlier?

_____

Calculate the mean without that outlier.

Without the outlier the average is

_____.

Use these boxes to show your work.

1 – 2.

3. The length of a rectangular carpet is 7 feet longer than its width. The area of the carpet is 800 square feet.

   What is the length of the carpet? What is the width?

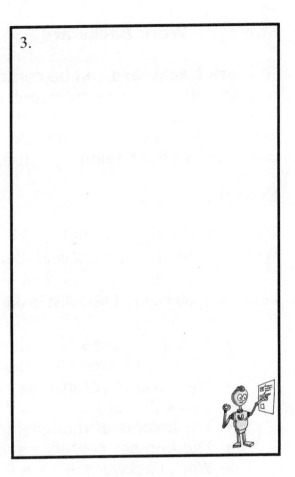

4. I'm thinking of two numbers whose product is 169 and whose difference is 0.

   What are the two numbers?

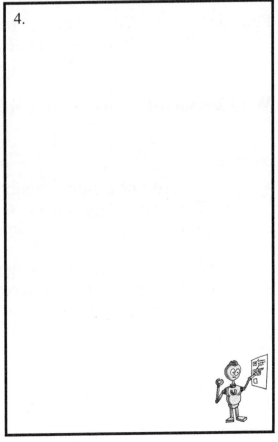

Lesson #3

Strategy: **Work Backward**

To **work backward** can be compared to using the rewind button on a remote control. This strategy is helpful when the order of events and the ending is known. Start at the end and do the steps in reverse order to find the beginning value.

Example:

The area of a right triangle is 18 sq. cm. The base is 4 cm long. What is the height of the triangle? Use this formula: Area = (b x h) ÷ 2

Find what you know. Plan what to do. Ask if the answer is reasonable.

- To find the area of a triangle, the base and height are multiplied. Then, that product is divided by two.
- The reverse (or inverse) of dividing in half is multiplying by two. (18 x 2 = 36)
- The inverse of multiplying (b x h) is dividing. 36 ÷ 4 = 9 cm
- The number sentence showing the formula is (4 x n) ÷ 2 = 18.
- Work backward → 18 x 2 = 36; 36 ÷ 4 = 9

Final answer: The height of the triangle is 9 cm, because 9 x 4 ÷ 2 = 18 cm. This answer is reasonable.

Work backward to solve the next four problems.

Use these boxes to show your work.

1. The area of a right triangle is 27 sq. cm. The base is 6 cm in length.

    What is the height of the triangle?
    Use this formula: Area = (b x h) ÷ 2.

1.

2. To find an average, add all the values, and then divide by the number of addends.

   The average, or mean, of which three percentages below, is approximately 92%?

   75%   88%   92%   95%   100%

2.

3. The fence around a rectangular field has a perimeter of 32 feet. The width of the field is 7 feet.

   Find the length of the field and its area.

3.

4. The area of an isosceles triangle is 33 sq. mm. The height is 11 mm.

   What is the base of the triangle?
   Use this formula: Area = $(b \times h) \div 2$.

4.

Lesson #4

1. The highest temperature ever recorded in the U.S. was 134°F in southwest California. This occurred in 1913. The lowest U.S. temperature ever recorded was -79.8°F in northern Alaska in 1971.

   What is the difference in these two temperatures?

2. The border around a rectangular field of sunflowers has a perimeter of 42 yards. The width of the field is 9 yards.

   Find the length and the area of the sunflower field.

Use these boxes to show your work.

1.

2.

3. Mount Etna in Sicily is one of the most active volcanoes on Earth. Mount Etna's height is close to 11,000 feet. Rocco wants to make a model of this volcano from clay.

If Rocco wants to make a scale model of the volcano, which of the following scales is most suitable? Use the box to explain your choice.

1 inch = 1,000 feet

or

1 inch = 100 feet

4. To find an average, add all the values, and then divide by the number of addends.

The average, or mean, of which three percentages below is approximately 78%?

65%   75%   85%   95%

3. _____

4.

Lesson #5

Strategy: **Look for a Pattern**

Sometimes a problem asks that a pattern be continued. A **pattern** is an idea that repeats. To write what comes next in a pattern, study the given information. Notice if a pattern is increasing or decreasing. Then identify the math operation(s) that is/are creating the pattern.

Example:

In order to change a fraction to a decimal, divide the numerator by the denominator. The quotient will either terminate (end) or repeat. The decimal equivalents in each fraction "family" create a pattern. Calculate the decimal equivalents to fill in the table. Memorize these values. (Use the pattern to help you.)

Find what you know. Plan what to do. Ask if the answer is reasonable.

- The denominator is always 12.
- Divide the numerator by the denominator.
- Some of the fractions can be simplified to lowest terms. Examples are $\frac{3}{12} = \frac{1}{4}$ and $\frac{6}{12} = \frac{1}{2}$.
- Noticing the pattern makes it easier to remember.

Fraction	Decimal
1/12	0.08$\overline{33}$
2/12	0.1$\overline{66}$
3/12	0.25
4/12	0.$\overline{33}$
5/12	0.41$\overline{66}$
6/12	0.50
7/12	0.58$\overline{33}$
8/12	0.$\overline{66}$
9/12	0.75
10/12	
11/12	
12/12	

Final answer: Continue to fill in chart: $\frac{10}{12} = 0.8333... = 0.8\overline{33}$

$\frac{11}{12} = 0.91666... = .019\overline{66}$, and $\frac{12}{12} = 1.00$.

The pattern occurs in sets of 3. In the first set of three, the first decimal ends in 33, the second in 66, and the third is equal to 1 quarter. In the second set of three, the first decimal ends in 33, the second in 66, and the third is equal to 2 quarters, and so on. The difference between each decimal equivalent is 0.08$\overline{33}$.

Look for the pattern in the decimal equivalents of each fraction "family" as you solve the next four problems. Continue to divide until the quotient terminates or begins to repeat.

12

Summer Solutions© Problem Solving   Level 7

Remember, in order to change a fraction to a decimal, divide the numerator by the denominator. The decimal equivalents in each fraction "family" create a pattern. Calculate the decimal equivalents to fill in each of the four tables below. Memorize these values. (Use the pattern to help you.)

Use these boxes to show your work.

1.

Fraction	Decimal
1/3	
2/3	
3/3	

2.

Fraction	Decimal
1/4	
2/4	
3/4	
4/4	

3.

Fraction	Decimal
1/5	
2/5	
3/5	
4/5	
5/5	

4.

Fraction	Decimal
1/6	
2/6	
3/6	
4/6	
5/6	
6/6	

Lesson #6

1. In this problem, identify the information that is STILL NEEDED in order to find the solution. If no information is needed, solve the problem.

   Mrs. Fields is bringing cans of soft drinks for the players to enjoy after softball practice.

   Will three 6-packs be enough for everyone?

   _____

   _____

   _____

2. The decimal equivalents in each fraction "family" create a pattern.

   Calculate the decimal equivalents to fill in the table. Memorize these values. (Use the pattern to help you.)

Fraction	Decimal
1/7	
2/7	
3/7	
4/7	
5/7	
6/7	
7/7	

Use these boxes to show your work.

1.

2.

3. The base of a parallelogram is 6 m longer than its height. The area of the parallelogram is 216 m².

   What is the measure of the base of the parallelogram? What is the height?

4. The walkway around a rectangular playground has a perimeter of 37 feet. The length of the playground is 10 feet.

   Find the width and the area of the playground.

Summer Solutions© Problem Solving   Level 7

Lesson #7

Strategy: **Use Logical Reasoning**

Logical reasoning is basically common sense. **Logical** means "sensible." **Reasoning** is "a way of thinking." Using logical thinking helps eliminate (remove) choices until the most reasonable answer (or conclusion) is left. Charts can help in organizing ideas.

Example:

Main Street is perpendicular to Forest Avenue. Forest Avenue runs in an east to west direction. Banneker Boulevard is east of Main St. and is also parallel to it. Make the drawing that best represents this description.

Find what you know. Plan what to do. Ask if the answer is reasonable.

- Perpendicular roads will form right angles.
- Parallel roads will be the same distance apart and will not intersect.
- East is the direction to the right when facing north.
- West is the direction to the left when facing north.

Final answer:

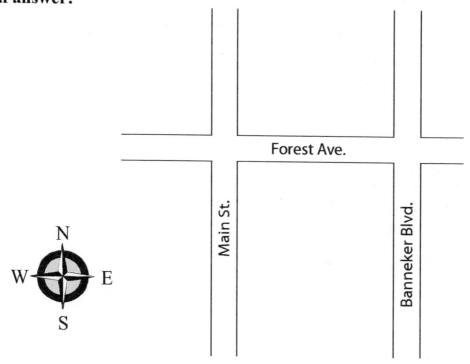

Use **logical reasoning** as you solve the next four problems.

16

Summer Solutions© Problem Solving **Level 7**

Use these boxes to show your work.

1. Interstate 79 (I-79) runs north/south through some of Pennsylvania. I-79 is perpendicular to I-80. I-376 is west of I-79 and runs parallel to I-79.

   Make the drawing that best represents this description.

2. A reasonable sum for $\frac{6}{7} + \frac{10}{11}$ would be 2. This is because $\frac{6}{7}$ is almost $\frac{7}{7}$ (or 1 whole) and $\frac{10}{11}$ is nearly $\frac{11}{11}$ (or one whole).

   What is a reasonable sum for the following problem? Use the box to explain your answer.

   $\frac{6}{7} + \frac{7}{8} + \frac{4}{5} + \frac{9}{10} =$ _____

3. Decide which statement is true. Then explain your reasoning in the box.

   A) All rectangles are squares.
   B) All squares are rectangles.

4. Eric's sister is a teenager. This year her age is a prime number. The year after next, her age will be a prime number again.

   What age is Eric's sister? Use the box to explain your answer.

Lesson #8

1. A reasonable product for 3.11 x 6.9 would be 21. This is because 3.11 rounds to 3 and 6.9 rounds to 7. The product of 3 and 7 is 21.

   What is a reasonable product for the following problem? Explain your answer.

   65.87 x 2.23 = _____

   _____

   _____

2. To calculate the mean of a set of numbers, add the values, and then divide by the number of addends.

   The average, or mean, of which two values below is $10.10?

   $7.20 $8.15 $15.00 $12.05

Use these boxes to show your work.

1.

2.

3. A regulation tennis court for singles matches has a perimeter of 210 feet. The length of the court is 78 feet.

Find the width of the court and the area of the court.

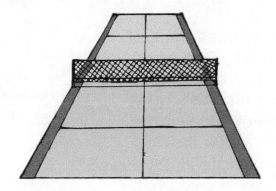

4. The decimal equivalents in each fraction "family" create a pattern.

Calculate the decimal equivalents to fill in the table. Memorize these values. (Use the pattern to help you.)

Fraction	Decimal
1/8	
2/8	
3/8	
4/8	
5/8	
6/8	
7/8	
8/8	

Lesson #9

Strategy: **Make an Organized List**

A math question may ask for a list of all the correct responses for a problem. **Using an organized list** is a way of arranging all the possibilities, so that none are overlooked or duplicated.

Example:

To calculate the number of possible outcomes, multiply the number of individual choices together. One spinner is divided into three equal sections: green, red, and blue. Another spinner is divided into 5 equal sections numbered one through five. Cassia spins the spinner. How many different outcomes are possible? What is the probability that her spin will result in red combined with any number?

Find what you know. Plan what to do. Ask if the answer is reasonable.

- There are 3 color choices and 5 number choices.
- Multiply the number of choices represented on each spinner to get the total possible outcomes (3 x 5 = 15). This is reasonable.
- The red choices are Red-1, Red-2, Red-3, Red-4, and Red-5.

Final answer: There are 15 different outcomes altogether. Five of them are red. The probability is five out of 15 ($\frac{5}{15}$ or $\frac{1}{3}$) that a player will get a number along with a red outcome.

Make an organized list to solve the next four problems.

Use these boxes to show your work.

1. Students at Camp Iroquois are being assigned an identification code. The code consists of one letter and one number.

   How many unique codes can the camp staff create using this system? An example has been given.

One Letter	One Number	Example
A through Z	0 through 9	B8

2. A cloth bag contains one marble of each of these colors: red, white, yellow, blue, and green. Manny places his hand in twice, each time drawing one marble without replacing it.

   How many different combinations can Manny choose?

   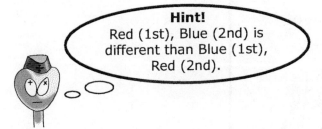

   Hint!
   Red (1st), Blue (2nd) is different than Blue (1st), Red (2nd).

3. The Spine Clinic is only open Friday morning. Patients can see either Dr. Backus or Dr. Tall. The doctors begin seeing patients every 20 minutes beginning at 8 a.m. The last appointment begins at 11:20 a.m.

   How many total appointment times can the Spine Clinic offer to patients on Friday morning?

4. Dennis is adding fruit snacks to his diet. He will have a morning and an afternoon snack. His choices are a banana, an apple, an orange, and a fruit roll-up.

   How many different combinations of 2 different fruits could Dennis have? (Morning-apple, afternoon-banana is different than morning-banana, afternoon-apple.)

Lesson #10

Use these boxes to show your work.

1. In a discount department store, bins filled with school supplies are marked "Choose any 3 for $2.25." These items can be chosen from the bins:

   - a notebook
   - a pack of filler paper
   - a set of markers

   How many different combinations of three items can be chosen? (Three notebooks is one choice.)

2. The decimal equivalents in each fraction "family" create a pattern.

   Calculate the decimal equivalents to fill in the table. Memorize these values. (Use the pattern to help you.)

Fraction	Decimal
1/9	
2/9	
3/9	
4/9	
5/9	
6/9	
7/9	
8/9	
9/9	

1.

2.

3. Ms. Marks wants to include more vegetables in her daily diet.

   If she chooses one green, leafy vegetable and one root, or tuberous, vegetable each day, how many different combinations can be created?

Green, Leafy	Root or Tuberous
Endive	Potato
Lettuce	Sweet Potato
Kale	Parsnip
Romaine	Carrot

   3.

4. The decimal equivalents in each fraction "family" create a pattern.

   Calculate the decimal equivalents to fill in the table. Memorize these values. (Use the pattern to help you.)

Fraction	Decimal
1/10	
2/10	
3/10	
4/10	
5/10	
6/10	
7/10	
8/10	
9/10	
10/10	

   4.

Lesson #11

Strategy: **Solve a Simpler Problem**

Each new school year there are more challenging ways to think about math. When a problem uses complicated numbers, try to **solve a simpler problem**. Using easier numbers keeps the focus on the steps needed to solve the problem.

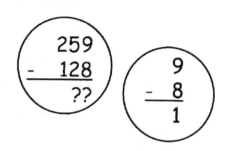

Example:

Pencils cost 15¢, while pens cost 25¢. Jon bought the same number of pencils as pens. If he spent $2, how many pencils and how many pens did Jon purchase?

Find what you know. Plan what to do. Ask if the answer is reasonable.

- If Jon bought one of each that would be simpler.
- One pencil and one pen cost 40¢ altogether.
- It is now easier to see that 5 x 40 equals $2.00.

Final answer: Five pencils cost 75¢ (15¢ x 5) and five pens cost $1.25 (25¢ x 5). The total is $2.00 ($1.25 + $0.75), so this is a sensible answer.

For the next four questions, **solve a similar, but simpler problem** first.

Use these boxes to show your work.

1. Bags of chips cost 50¢, while soft drinks cost 75¢. Aimee bought the same number of chips as soft drinks.

   If she spent $5, how many bags of chips and how many soft drinks did Aimee purchase?

1.

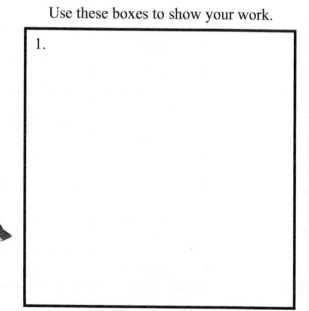

2. Look at each algebraic expression in the chart. Match it to its simplest form. Choose from the expressions below.

$3n + \frac{1}{2}$ $n^2 + 6$

$n^3 + 8$ $2n + 6$

Simplest Form	Algebraic Expression
	$n + n + n + (8/16)$
	$n \times n \times n + (2^3)$
	$n + n + (2 \times 3)$

3 – 4. Lynda asked her friends to record the amount of their weekly allowance.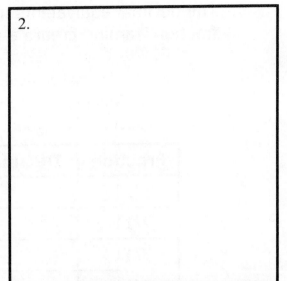

The results are in the chart.

Name	Weekly Allowance
Ben	$3.00
Allie	$4.50
Jon	$10.00
Megan	$5.00
Kenny	$6.50

What is the mean weekly allowance?

_____

Which four friends have an average allowance of less than $5?

_____

_____

The outlier in this set of data is

_____.

Lesson #12

1. The decimal equivalents in each fraction "family" create a pattern.

   Calculate the decimal equivalents to fill in the table. Memorize these values. (Use the pattern to help you.)

Fraction	Decimal
1/11	
2/11	
3/11	
4/11	
5/11	
6/11	
7/11	
8/11	
9/11	
10/11	
11/11	

2. A reasonable product for 2.21 x 3.8 would be 8. This is because 2.21 rounds to 2 and 3.8 rounds to 4. The product of 2 and 4 is 8.

   What is a reasonable product for the following problem? Explain your answer.

   211.75 x 7.82 = ___

   _____

   _____

Use these boxes to show your work.

3. Look at each algebraic expression in the chart. Match it to its simplest form. Choose from the expressions below.

$n^2 + 6$ $3n + 9$
$n^2 + 9$ $2n + 6$

Simplest Form	Algebraic Expression
	$n + n + (18/3)$
	$n \times n + (3^2)$
	$3(n + 3)$

4. Sabine is a teenager. This year her age is a multiple of three. Last year her age was a multiple of seven. Next year her age is a perfect square.

How old is Sabine? Explain your answer.

Lesson #13

Strategy: **Use a Table / Make a Table**

Writing ideas in some type of **table** (or chart) helps to organize data. A simple table is T-shaped. Other tables have rows, columns, and labels. Graphs can help organize data, too. At times, patterns become visible in the data by using these organizers.

Example:

Find the mean, median, mode, and range of the data represented in this graph.

Mean: _____   Median: _____

Mode: _____   Range: _____

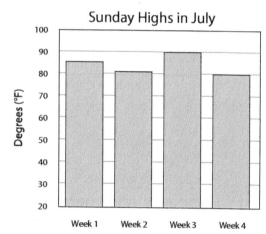

Find what you know. Plan what to do.
Ask if the answer is reasonable.

- The mean, or average, is found by adding all the values, and then dividing by the number of addends.
- The range is the difference between the largest and smallest numbers.
- The median is the middle number, after the data has been arranged from lowest to highest.
- The mode is the number that appears most often in the data.

Final answer: The mean is 84 degrees. The median temperature is 83 degrees. These temperatures have no mode. The range of the temperatures is 10 degrees.

For the next four problems, **use a table or graph**.

1 – 2. How many total minutes did Terri exercise this week? Find the mean, median, mode, and range of the data represented in this line graph.

Mean: _____   Median: _____

Mode: _____   Range: _____

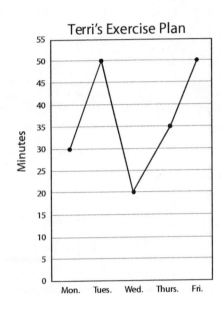

3 – 4. Carey's brother owed $10,000 on his college loan. He has made payments of $250 for 7 months.

How much does Carey's brother still owe?

_____

If Carey's brother is not charged any additional interest or fees, how many more months will he need to pay this debt off at this rate?

_____

That is (more / less) than 3 years.

Use these boxes to show your work.

1 – 2.

3 – 4.

Month	Balance Owed

Lesson #14

1. To calculate the number of possible outcomes, multiply the number of individual choices together. For example, if 3 flavors of ice cream are being offered with 2 flavors of topping, there are a total of 3 x 2, or 6, possible combinations of ice cream with topping. Use that skill to make sure a list of possible outcomes is complete.

   A hospital cafeteria has a list of breakfast choices. Luc chooses one item from each group listed.

   How many different breakfast meals of three items can be created?

Whole Grain	Fruit	Milk
Oatmeal	Blueberries	Skim
Quinoa	Strawberries	Soy
Shredded wheat	Bananas	Rice
Toast (multigrain)		

2. Will's brother is in his twenties. Last year his age was a prime number. Next year his age will be a perfect square.

   What age is Will's brother? Give one interesting math fact about that number.

Use these boxes to show your work.

1.

2.

3 – 4. How much time did Randy spend on math problem solving during the 5-week period?

Find the mean, median, mode, and range of the data represented in this graph.

Mean: _____   Median: _____

Mode: _____   Range: _____

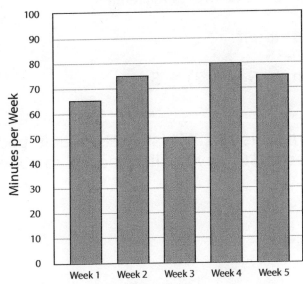

3 – 4.

Lesson #15

Strategy: **Write an Equation**

Grammatically speaking, a sentence consists of a subject and a predicate. An **equation** (number sentence) is made up of numbers and math symbols (+ - ÷ × = < >). The equal sign becomes the verb. To use this strategy, turn the words of a problem into numbers and symbols.

Example:

Shari purchased a meal that was $5 more than twice the cost of Angie's meal. Angie's meal cost $3. What was the cost of Shari's meal?

Find what you know. Plan what to do. Ask if the answer is reasonable.

- Twice means *two times the amount*.
- The words *$5 more* become + 5.
- The equation for this problem is $S = 2(3) + 5$.

Final answer: The cost of Shari's meal is $11. This is reasonable because Angie's meal doubled is $6. Five dollars added to six dollars is $11.

Write an equation for the next four problems.

Use these boxes to show your work.

1. Monique bought swim goggles that were $8 more than twice the cost of a bottle of sunscreen. The swim goggles cost $21.98.

   What was the cost of the sunscreen?

2. Ji purchased a video game that was $20 more than twice the cost of a DVD. The video game cost $45.

What was the cost of the DVD?

3. Choose the equation below that represents these words:

The product of a number and 7 is equal to one less than twenty-two.

Then, find the value of n.

A) 7 + n = 1 − 22
B) 7n = 1 − 22
C) 7n = 22 − 1
D) 7n + 1 − 22

4. Choose the expression below that represents these words:

Twelve less than the square of ten.

A) $12 - 10^2$
B) $10^2 - 12$
C) $(12 - 10)^2$

Lesson #16

1. The Bureau of Motor Vehicles in Florida has decided on the license plate design for the coming year. The design will show two letters followed by 3 numbers.

   How many unique license plates can be created using this plan? An example has been given. Write two additional examples.

2 Letters	3 Numbers	Example
A through Z	0 through 9	SQ643

   Use these boxes to show your work.

   1.

2. The rules of a board game tell players to roll two number cubes, subtract the numbers on the cubes, and move that many spaces on the board. One cube is gray and the other is white. The cubes are numbered 1 through 6.

   Use the table to record all the possible differences. A few have been done for you.

   How many different ways can a player get a difference of 1?

   _____

   What is the probability of getting a difference of one?

   _____

   2.

	1	2	3	4	5	6
1	0				4	
2		0		2		
3		1				
4						
5	4					
6						

3. Look at each algebraic expression in the chart. Match it to its simplest form. Choose from the expressions below.

8m + 1   4m² + 4m + 24

12m + 16   m³ + 10

Simplest Form	Algebraic Expression
	4(m^2 + m + 6)
	4(2m + m + 4)
	4(2m + 1/4)

4. Yoder's Farm raises chickens and sheep. They have the same number of each animal. Casey counted 48 legs when all the animals were roaming about one day.

How many sheep and chickens are raised at Yoder's Farm?

Lesson #17

Strategy: **Make a Model**

A **model** can be a drawing or a homemade object that helps the words of a problem take on a physical shape. Some of these objects might include coins, paper clips, paper (for folding), or cubes, to name a few.

Example:

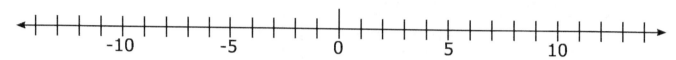

Here is a way to understand addition and subtraction of integers by using a hiking model.

- The start location is the first number.
- Next, the math operation tells the hiker which direction to face (+ → right/east and – → left/west).
- Then, the second number indicates if the hiker should walk forward (positive integer) or backward (negative integer).
- Finally, the hiker reaches the destination (answer).

The following hiking examples attempt to model the number problem.

1) -8 + -3 → Start at -8, face right (+), and walk backward 3 steps (-3). [Answer: -11]

2) -3 - -4 → Start at -3, face left (-), and walk backward 4 steps (-4). [Answer: +1]

Use the integer number line to solve 9 + -12 = _____.

Find what you know. Plan what to do. Ask if the answer is reasonable.

- Start at positive 9.
- To add, face right or east.
- The second integer is negative, so the hiker walks backward 12 steps (-12).

Final answer: The hiker ends at -3.

Make a model to help you understand the next four items.

1. Use the integer number line on the opposite page to solve 7 + -15 = _____.

2. Use the integer number line on the opposite page to solve -10 + 15 = _____.

Use these boxes to show your work.

3. This drawing is a model of finding the mean, or average, of a set of numbers. In the first drawing there are three stacks of 5, 1, and 3 coins. If the stacks were adjusted to make them equal, there would be 3 coins in each stack. An equation for this model is 5 + 1 + 3 = 9. Nine total coins divided into 3 equal sections would amount to 3 coins in each stack (9 ÷ 3 = 3).

3.

Make a similar model to demonstrate finding the mean of 5 stacks of 9, 2, 6, 1, and 2 cubes.

4.

4. Make a model similar to the one in Item 3 to demonstrate finding the mean of 3 stacks of 4, 7, and 4 coins.

37

Lesson #18

1. Felipe sold a portable media player for $15 more than twice the cost of a CD. The media player cost $46.98.

   What was the cost of the CD?

2. Fill in the solution column below with the correct answer.

   The solution choices are 12, 10, and 14.

   An answer will be used more than once.

   Follow the rules for the order of operations.

   Order of Operations:
   1. Solve inside brackets or parentheses.
   2. Calculate the exponents or roots.
   3. Multiply or divide from left to right.
   4. Add or subtract from left to right.

Expression	Solution
2(3 + 4)	
2 × 3 + 4	
$2^3 + 4$	
2 · (3 + 4)	

Use these boxes to show your work.

1.

2.

3. Look at each algebraic expression in the chart. Match it to its simplest form. Choose from the expressions below.

7x   3 + 4x

5 + 2x   5x + 2

Simplest Form	Algebraic Expression
	3x + 2x + 2x
	3 + 2x + 2
	3 + 2x + 2x

4. Murray's Scrap Metal recycles old bike parts. On Friday, a delivery truck brought in the same number of bicycles as tricycles. Patrick Murray counted 45 tires.

How many bicycles and how many tricycles were delivered on Friday?

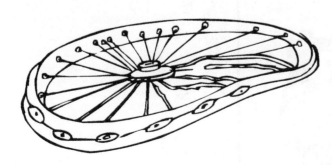

Lesson #19

Strategy: **Draw a Picture or Diagram**

There's a saying: "A picture is worth a thousand words." Turning the words of a math problem into a **picture or diagram** makes it easier to imagine the problem.

Example:

The perimeter of a rectangle is 34 cm. The length of one side is 9 cm. What is the width of the rectangle?

Find what you know. Plan what to do.
Ask if the answer is reasonable.

- A rectangle has two pairs of parallel sides.
- The perimeter is the distance around the rectangle.
- If one side measures 9 cm, the opposite side is also 9 cm.
- Imagine the rectangle; draw it and label what you know.
- The total perimeter (34) minus 18 (9 + 9) equals 16.
- The remaining amount must be divided equally between the remaining two sides. (16 ÷ 2 = 8)

Final answer: The rectangle has a width of 8 cm. This is reasonable because 9 + 9 + 8 + 8 = 34.

Solve the next four problems by **drawing a picture**.

Use these boxes to show your work.

1. The perimeter of a rectangle is 443.54 inches. The width of one side is 108.24 inches.

    What is the length of the rectangle?

2. The perimeter of a rectangle is 112 cm. The length of one side is 27 cm.

   What is the width of the rectangle?

3. In football, the offensive players are given four downs, or plays, to move the ball ten yards towards their opponent's end zone. If the players succeed, it is called "getting a first down." They then keep possession of the football and get another four downs to move the ball the same distance.

   Read the following recap of several minutes from the Cougars' game. Did the team get a first down? Tell how many yards the ball was moved and if this was a gain (+) or loss (-) in yards.

First down	gain of 5 yards
Second down	loss of 8 yards
Third down	gain of 9 yards
Fourth down	gain of 2 yards

4. Joy is taking care of her neighbor's plants for two weeks. The lily needs to be watered every 3rd day; the violet needs to be watered every 4th day, and the spider plant needs to be watered every 6th day.

   On which day will all of the plants need to be watered?

    Make a diagram of the days in two weeks. Then circle each type of plant's watering days in a different color.

Lesson #20

Use these boxes to show your work.

1. When the sun comes up each morning, temperatures gradually increase until about midday. It is 71 degrees at 9:30 a.m.

   If the temperature increases one degree every 5 minutes, at what time will it be 80 degrees?

2. The counselors asked a group of campers to vote for their favorite activity.

   Complete the circle graph by creating sections that represent the results of the survey. Give the graph a title.

   - 50% Swimming
   - 10% Biking
   - 30% Tennis
   - 10% Golfing

Title:_____

3. Mrs. Walker priced her special chocolate cake at $6 more than twice the cost of a loaf of her multigrain bread. The cake sold for $13.

What was the cost of Mrs. Walker's bread per loaf?

4. Fill in the solution column with the correct answer. The solution choices are 21, 49, 53, 80, 128, and 148.

Follow the rules for the order of operations.

Expression	Solution
$52 + 7 \cdot 4$	
$5^2 + 7 \times 4$	
$(5^2 + 7) \times 4$	
$(5 + 7)^2 + 4$	
$5(2 + 7) + 4$	

3.

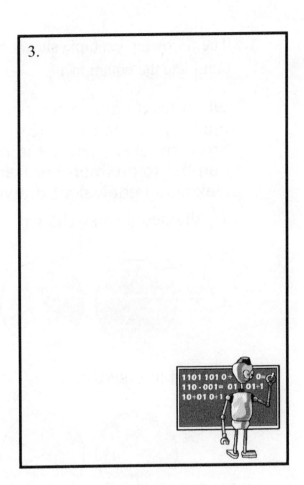

4.

Lesson #21

1. The following example shows how to complete the equation $1\frac{1}{3} \div \frac{1}{6}$.

   When dividing by a fraction, multiply by the reciprocal. In this problem, also change the mixed number to an improper fraction. Making an equivalent drawing of $1\frac{1}{3}$ divided into sixths will help.

   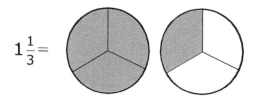

   divided into sixths...

   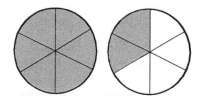

   So,

   $1\frac{1}{3} \div \frac{1}{6} \rightarrow \frac{4}{3} \times \frac{6}{1} = \frac{24}{3} = 8$.

   This means if you divided (or re-cut) $1\frac{1}{3}$ into sixths, there would be 8 of that size.

   Make a similar model to demonstrate $2\frac{1}{2} \div \frac{1}{4}$.

2. Water freezes at 32°F. What temperature is 50°F below freezing?

Use these boxes to show your work.

1.

2.

3. Read the following recap of several minutes from the Bears' game. Did the team get a first down? Tell how many yards the ball was moved and if this was a gain (+) or loss (-) in yards.

   First down   gain of 8 yards
   Second down   loss of 9 yards
   Third down   loss of 7 yards
   Fourth down   gain of 19 yards

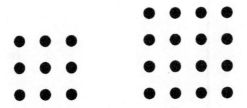

4. Look at the diagrams above.

   The product of a number and itself is called the square of the number. It is clear that the area of the shapes above form a square design.

   Can 64 square units be arranged in a square shape?

   _____

   Use the graph paper in the box to demonstrate whether or not 64 is a perfect square.

   What is the length of each side of the square?

   _____

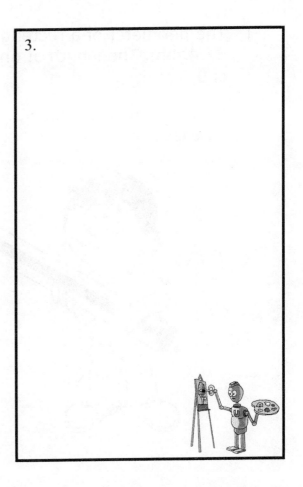

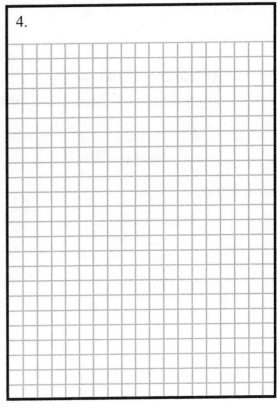

Lesson #22

1. The perimeter of a rectangle is 32.4 mm. The length of one side is 9.3 mm.

   What is the width and the area of the rectangle?

2. To calculate the mean (average) of a set of numbers, find the sum of the set, and then divide by the number of addends.

   The average, or mean, of which three values below is 36?

   24   52   48   61   36

3. Solve $3\frac{3}{4} \div \frac{1}{8}$.

   Make a model similar to the one in Lesson #21 to help you.

4. A basketball game ended at 9:38 p.m. The first quarter was 22 minutes long. The second quarter was 29 minutes long. The half-time break lasted 14 minutes. The second half was an hour and 3 minutes long.

   At what time did the basketball game begin?

3.

4.

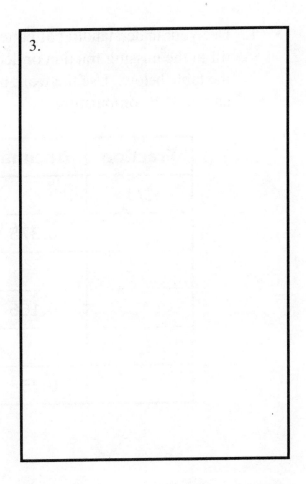

Lesson #23

1. Use your understanding of patterns to fill in the missing fraction or decimal in the table below. Use the work box for any needed computation.

Fraction	Decimal
1/3	
	0.375
1/9	
	0.1$\overline{66}$
1/7	
	0.75

Use these boxes to show your work.

1.

2. The perimeter of a rectangle is 12 ft. The length of one side is $4\frac{3}{4}$ ft.

What is the width of the rectangle?

Which choice below is a reasonable estimate for the area of the rectangle? Circle it.

25 sq. ft.   5 sq. ft.

2.

3. Solve $2\frac{2}{3} \div \frac{1}{6}$.

   Make a model similar to the one in Lesson #21 to help you.

4. Make a model similar to the one in Lesson #17 to demonstrate finding the mean of four stacks of 8, 4, 3, and 9 coins.

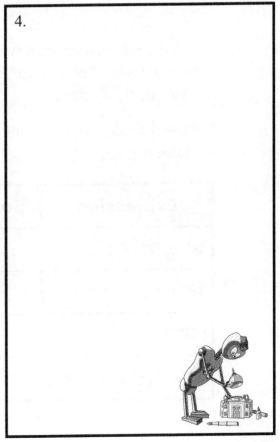

Lesson #24

Use these boxes to show your work.

1. When the sun comes up each morning, temperatures gradually increase until about midday. It is 79°F at 11 a.m.

   If the temperature increases one degree every 10 minutes, what will the temperature be at noon?

2. Fill in the solution column with the correct answer. The solution choices are 12, 6, 4, 3, and 2.

   Follow the rules for the order of operations.

Expression	Solution
48 ÷ 8 − 4 × 1	
48 ÷ (8 − 4) × 1	
(48 ÷ 8 − 4) + 1	
48 ÷ [(8 + 4) × 1]	

1.

2.

3. Solve $2\frac{1}{4} \div \frac{1}{8}$.

Make a model similar to the one in Lesson #21 to help you.

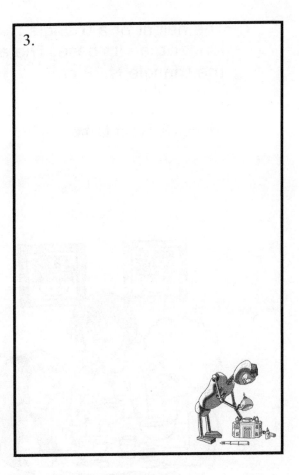

4. Read the following recap of several minutes from the Sharks' game. Did the team get a first down? Tell how many yards the ball was moved and if this was a gain (+) or loss (-) in yards.

First down	loss of 9 yards
Second down	loss of 9 yards
Third down	loss of 9 yards
Fourth down	gain of 48 yards

Lesson #25

1. The height of a triangle is 4 cm longer than its base. The area of the triangle is 30 cm².

   What is the measure of the base of the triangle? What is the height?

2. Grandma Mary wants to serve dinner at 5:30 p.m. The directions for a roast indicate to allow 20 minutes for every pound. After removing it from the oven, the roast should "rest" for 15 minutes. This creates a tasty juice. She'll need 10 minutes to carve the roast and arrange it on a tray. The roast weighs four pounds.

   At what time should Grandma Mary put the roast in the oven?

Use these boxes to show your work.

Summer Solutions© Problem Solving   Level 7

3. In this problem, identify the information that is STILL NEEDED in order to find the solution. If no information is needed, solve the problem.

   A group of 4 friends earned $64 weeding and mowing lawns one day in their neighborhood.

   If the friends agreed to split the money evenly, how much did each person earn?

   _____

   _____

   _____

3.

4. Which polygon is being described below? Use the box to explain your reasoning.

   A quadrilateral with all sides congruent and all angles congruent is a

   _____.

4.

Lesson #26

1. The width of the state of Colorado is 380 miles. The length of the state is 280 miles. Lena wants to draw the map of Colorado to scale for her state book report. The paper Lena will use is standard-sized 8.5 by 11 inches.

   Which of the following scales make the best sense for Lena to use? Use the box to explain your choice.

   1 inch = 10 miles

   or

   1 inch = 40 miles

2. A map of downtown Cleveland, Ohio shows five streets. Chester, Euclid, and Carnegie Avenues are parallel to each other for several miles. Euclid Avenue is south of Chester Avenue and north of Carnegie Avenue. The three streets are intersected by East 55th and East 79th Streets. East 79th is east of 55th Street. Both streets are perpendicular to the avenues.

   Make the drawing that best represents this description.

Use these boxes to show your work.

3. A local restaurant is featuring these specials. Customers choose one entrée (main dish) and two different sides for $7.99.

   How many combinations can be ordered? Give two different examples.

Main Entrée	Sides
Grilled Chicken	Hash Browns
Tilapia (fish)	Sweet Potato
Veggie Burger	Applesauce
Bean Burger	Salad
Pasta with Sauce	Green Beans

4. Look at each algebraic expression in the chart. Match it to its simplest form. Choose from the expressions below.

   2x + 5   5x + 2
   4x + 3   5x + 3

Simplest Form	Algebraic Expression
	x + 1 + 3x + 2
	x + x + 3x + 2
	2 + 1 + 3x + 2x

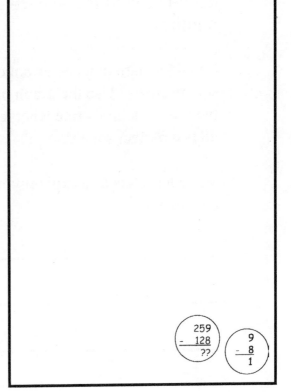

Lesson #27

Use these boxes to show your work.

1. The average person can walk 3 miles in an hour. The average person can bicycle 10 miles in an hour.

   Jo and Anna are average people. If Anna bikes 40 miles in the same amount of time that Jo walks, how many miles has Jo walked?

2. The product of a number and itself is called the square of the number.

   Can 48 square units be arranged in a square shape? Use the graph paper in the box to demonstrate whether or not 48 is a perfect square.

   Write a sentence to explain your conclusion.

   _____

   _____

   _____

3. Use the integer number line to solve -8 – -10 = _____.

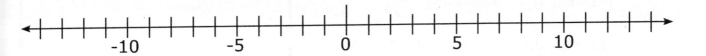

4. Choose the equation below that represents these words:

Twice the sum of five and a number is equal to the product of six and three.

Then, find the value of *n*.

A) 2 x 5 + *n* = 6 x 3

B) 2 + 5 x *n* = 6 + 3

C) 2(5 + *n*) = 6 x 3

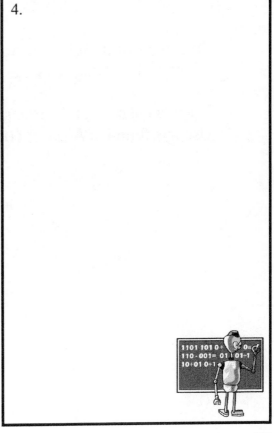

Lesson #28

1. I'm thinking of two numbers whose difference is 0 and whose product is 225.

   What are the two numbers?

2. The area of a right triangle is 42 m². The height is 12 m.

   What is the base of the triangle?
   Use this formula: Area = (b x h) ÷ 2.

3. The circumference of a circle is approximately three times the size of the diameter. The formula for calculating the circumference is $C = \pi d$ or $2\pi r$ ($\pi \approx 3.14$).

Use estimation skills to fill in the missing values in the chart. Choose from the measurements below.

314 ft.   240 ft.   349 ft.   292 ft.

Diameter	Circumference
93 ft.	
111 ft.	
100 ft.	

4. Which solid is being described below? Explain your reasoning.

A solid figure having two congruent circular bases connected by a curved surface is

a _____.

_____

_____

_____

Draw the figure that best represents this description.

Lesson #29

1. The circumference of a circle is approximately three times the size of the diameter. The formula for calculating the circumference is C = πd or 2πr (π ≈ 3.14). Remember, the diameter is twice the length of a radius.

   Use estimation skills to fill in the missing values in the chart. Choose from the measurements below.

   88 cm   60 cm   69 cm   57 cm

Radius	Circumference
14 cm	
9 cm	
11 cm	

   Use these boxes to show your work.

2. Use your understanding of patterns to fill in the missing fraction or decimal in the table below. Use the work box for any needed computation.

Fraction	Decimal
2/3	
	0.875
7/9	
	0.8$\overline{33}$
4/7	
	0.40

3. Look at each algebraic expression in the chart. Match it to its simplest form. Choose from the expressions below.

4x + 1   4x⁴
3x + 3   8x + 4

Simplest Form	Algebraic Expression
	4x · x · x · x
	x + x + 2x + 1
	4(2x + 1)

4. The rules of a board game tell players to shake and roll two number cubes, multiply the numbers on the cubes, and move that many spaces on the board. One cube is red and the other is blue. Each cube is numbered 1 through 6.

How many different outcomes are possible?

Hint!
Blue 6 Red 3 is a different outcome than Red 6 Blue 3. The product of 18 (3 x 6) was arrived at two different ways. Making a chart could prove helpful.

Lesson #30

1. At the Mitchell family reunion, the ages of the children were tallied. The percentages were calculated.

   Complete the circle graph by creating sections that represent the percentages below. Give the graph a title.

   - 25% Ages 9-12
   - 10% Ages 6-8
   - 25% Ages 3-5
   - 40% Ages 0-2

2. Make a model similar to the one in Lesson #17 to demonstrate finding the mean of 6 stacks of 2, 4, 8, 9, 3, and 4 cubes.

Use these boxes to show your work.

1. Title:_____

2.

3. These two rectangles are similar. The first rectangle has a length of 1.8 inches and a width of 1 inch. Each side of the second rectangle has been reduced by 50%.

What are the dimensions of the second rectangle?

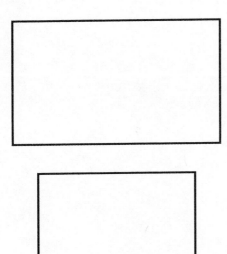

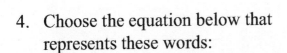

4. Choose the equation below that represents these words:

One more than the product of seven and a number is equal to one less than four squared.

Then, find the value of n.

A) $1 \times 7 + n = 1 - 4^2$

B) $7n + 1 = 4^2 - 1$

C) $1 + 7n = 1 - 4^2$

Level 7

Problem Solving

Help Pages

Help Pages

Problem Solving

Strategies

Guess and Check

For the **guess and check** strategy, make a guess. See if it fits all the clues. If it does, the item is solved. If it does not, let that idea improve the next guess.

Work Backward

This strategy is helpful when the order of events and the ending is known. Start at the end and do the steps in reverse order to find the beginning value.

Look for a Pattern

A pattern is an idea that repeats. To write what comes next in a pattern, study the given information. Notice if the numbers in a pattern are increasing or decreasing. Then identify the math operation(s) that is/are creating the pattern.

Use Logical Reasoning

Logical reasoning is basically common sense. **Logical** means "sensible." **Reasoning** is "a way of thinking." Using logical thinking helps eliminate (remove) choices until the most reasonable answer (or conclusion) is left. Charts can help in organizing ideas.

Make an Organized List

A math question may ask for a list of *all the correct responses* for a problem. Using an **organized list** is a way of arranging all the possibilities, so that none are overlooked or duplicated (repeated).

Solve a Simpler Problem

When a problem uses complicated numbers, try to **solve a simpler problem**. Using easier numbers keeps the focus on the steps needed to solve the problem.

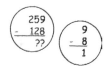

Use a Table/Make a Table

Writing ideas in some type of table (or chart) helps to organize data. A simple table is T-shaped. Other tables have rows, columns, and labels. Graphs can help organize data, too. At times, patterns become visible in the data by **using a table**.

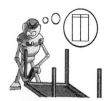

Write an Equation

An **equation** (number sentence) is made up of numbers and math symbols (+ − ÷ × = < >). The equal sign becomes the verb. To use this strategy, turn the words of a problem into numbers and symbols.

Help Pages

Problem Solving

Strategies (continued)

Make a Model
A model can be a drawing or a homemade object that helps the words of a problem take on a physical shape. Some of these objects might include coins, paper clips, paper for folding, or cubes, to name a few.

Draw a Picture
Turning the words of a math problem into a picture or diagram makes it easier to imagine the problem.

Vocabulary

Arithmetic

Equation — a math "sentence" that uses numbers, math symbols, and an "=" sign.

Expression — a mathematical phrase formed from numbers, variables, and operation symbols.

Integers — the set of whole numbers, positive or negative, and zero.

Variable — a symbol or letter that can stand for a number in an expression or equation.

Factors and Multiples

Composite Number — a number with more than 2 factors.
Example: 10 has factors of 1, 2, 5, and 10. Ten is a composite number.

Factors — are multiplied together to get a product. Example: 2 and 3 are factors of 6.

Greatest Common Factor (GCF) — the highest factor that 2 numbers have in common.
Example: The factors of 6 are 1, 2, **3**, and 6. The factors of 9 are 1, **3**, and 9. The GCF of 6 and 9 is 3.

Least Common Multiple (LCM) — the smallest multiple that 2 numbers have in common.
Example: Multiples of 3 are 3, 6, 9, **12**, 15... Multiples of 4 are 4, 8, **12**, 16... The LCM of 3 and 4 is 12.

Multiples — can be evenly divided by a number. Example: 5, 10, 15, and 20 are multiples of 5.

Perfect Square — the product of a number multiplied by itself; 16 is a perfect square because 4 × 4 = 16.

Prime Number — a number with exactly 2 factors (the number itself & 1).
Example: 7 has factors of 1 and 7. Seven is a prime number.

Fractions and Decimals

Improper Fraction — a fraction in which the numerator is larger than the denominator. Example: $\frac{9}{4}$

Mixed Number — the sum of a whole number and a fraction. Example: $5\frac{1}{4}$

Help Pages

Vocabulary

Fractions and Decimals (continued)

Reciprocal — a fraction in which the numerator and denominator are interchanged. The product of a fraction and its reciprocal is always 1.

Example: The reciprocal of $\frac{3}{5}$ is $\frac{5}{3}$. $\frac{3}{5} \times \frac{5}{3} = \frac{15}{15} = 1$

Repeating Decimal — a decimal which has a number or a series of numbers continuing on and on.
Example: $2.\overline{33}$, $4.\overline{15}$, $7.12\overline{55}$, etc.

Geometry

Acute Angle — an angle measuring less than 90°.

Circumference — the distance around the outside of a circle.

Congruent — figures with the same shape and the same size.

Obtuse Angle — an angle measuring more than 90°.

Reflection — a flip; when a figure is flipped over a given line.

Right Angle — an angle measuring exactly 90°.

Rotation — a turn; when a figure rotates on a point.

Similar — figures with the same shape but different sizes.

Straight Angle — an angle measuring exactly 180°.

Transformation — a change in the position of a geometric figure or shape.

Translation — a slide; sliding or moving every point in a figure the same distance in the same direction.

Volume — the number of cubic units needed to fill a solid.

Geometry — Polygons

Number of Sides		Name	Number of Sides		Name
3	△	Triangle	7	⬡	Heptagon
4	□	Quadrilateral	8	⬡	Octagon
5	⬠	Pentagon	9	⬡	Nonagon
6	⬡	Hexagon	10	⬡	Decagon

Parallelogram — a four-sided polygon with two pairs of parallel sides of equal length.

Trapezoid — a four-sided polygon with one set of parallel sides.

Help Pages

Vocabulary

Geometry — 3-Dimensional Shapes

Cube — a shape with six square faces.

Rectangular prism — a six-sided shape with rectangular faces.

Geometry — Triangles

Equilateral — a triangle with all 3 sides having equal length.

Isosceles — a triangle with 2 sides having equal length.

Scalene — a triangle with none of the sides having equal length.

Probability

Probability — the likelihood of an event occurring.

Outcome — a possible result that can occur in a situation.

Statistics

Mean — the average of a group of numbers. The mean is found by finding the sum of a group of numbers and then dividing the sum by the number of members in the group (addends).

Example: The average of 12, 18, 26, 17, and 22 is **19**. $\frac{12+18+26+17+22}{5} = \frac{95}{5} = 19$

Median — the middle value in a group of numbers. The median is found by listing the numbers in order from least to greatest, and finding the one that is in the middle of the list. If there is an even number of members in the group, the median is the average of the two middle numbers.
Example: The median of 14, 17, 24, 11, and 26 is **17**.   11, 14, 17, 24, 26

The median of 77, 93, 85, 95, 70, and 81 is **83**.   70, 77, 81, 85, 93, 95 $\frac{81+85}{2} = 83$

Mode — the number that occurs most often in a group of numbers. The mode is found by counting how many times each number occurs in the list. The number that occurs more than any other is the mode. Some groups of numbers have more than one mode.
Example: The mode of 77, 93, 85, 93, 77, 81, 93, and 71 is **93**.   (93 occurs more than the others.)

Range — the difference between the greatest and the least in a set of numbers.

Ratio and Proportion

Proportion — a statement that two ratios (or fractions) are equal. Example: $\frac{1}{2} = \frac{3}{6}$

Ratio — a comparison of two numbers by division; a ratio looks like a fraction. Example: $\frac{1}{2}$

Percent (%) — the ratio of any number to 100. Example: 14% means 14 out of 100, or $\frac{14}{100}$.

Help Pages

Vocabulary

Measurement — Relationships

Volume	Distance
3 teaspoons in a tablespoon	36 inches in a yard
8 ounces in a cup	1760 yards in a mile
2 cups in a pint	5280 feet in a mile
2 pints in a quart	100 centimeters in a meter
4 quarts in a gallon	1000 millimeters in a meter

Weight	Temperature
16 ounces in a pound	0° Celsius – Freezing Point
	100°Celsius – Boiling Point
2000 pounds in a ton	32°Fahrenheit – Freezing Point
	212°Fahrenheit – Boiling Point

Time

24 hours in a day	10 years in a decade
365 days in a year	100 years in a century

Place Value

Factors

The **Greatest Common Factor (GCF)** is the largest factor that 2 numbers have in common.

Example: Find the Greatest Common Factor of 32 and 40.

The factors of 32 are 1, 2, 4, 8, 16, 32
The factors of 40 are 1, 2, 4, 5, 8, 10, 20, 40

1. List the factors of each number.
2. Find the largest number that is in both lists.

The GCF of 32 and 40 is **8**.

Exponents

An **exponent** is a small number to the upper right of another number (the base). Exponents are used to show that the base is a repeated factor.

Example: 2^4 is read "two to the fourth power."

base ⟶ 2^4 ⟵ exponent

The base (2) is a factor multiple times.

The exponent (4) tells how many times the base is a factor.

$2^4 = 2 \times 2 \times 2 \times 2 = 16$

Example: 9^3 is read "nine to the third power" and means $9 \times 9 \times 9 = 729$

Summer Solutions© Problem Solving — Level 7

Help Pages

Solved Examples

Expressions

An **expression** is a number, a variable, or any combination of these, along with operation signs (+, -, ×, ÷) and grouping symbols. An expression never includes an equal sign.
Five examples of expressions are 5, x, $(x+5)$, $(3x+5)$, and $(3x^2+5)$.
To **evaluate an expression** means to calculate its value using specific variable values.
Example: Evaluate $2x + 3y + 5$ when $x = 2$ and $y = 3$.

$$2(2) + 3(3) + 5 = ?$$
$$4 + 9 + 5 = ?$$
$$13 + 5 = 18$$

1. To evaluate, put the values of x and y into the expression.
2. Use the rules for integers to calculate the value of the expression.

The expression has a value of 18.

Example: Find the value of $\frac{xy}{3} + 2$ when $x = 6$ and $y = 4$.

$$\frac{6(4)}{3} + 2 = ?$$
$$\frac{24}{3} + 2 = ?$$
$$8 + 2 = 10$$

The expression has a value of 10.

When evaluating a numerical expression containing multiple operations, use a set of rules called the **Order of Operations**. The Order of Operations determines which operations, and in which order, they should be performed. (Which operation should be done first, second, etc.)
The Order of Operations is as follows:

1. Parentheses
2. Exponents
3. Multiplication/Division (left to right in the order that they occur)
4. Addition/Subtraction (left to right in the order that they occur)

If parentheses are enclosed within other parentheses, work from the inside out.
To remember the order, use the mnemonic device "Please Excuse My Dear Aunt Sally."

Help Pages

Solved Examples

Expressions (continued)

Use the following examples to help you understand how to use the Order of Operations.

Example: $2 + 6 \cdot 5$

To evaluate this expression, work through the steps using the Order of Operations.
1. Since there are no parentheses or exponents in the expression, skip steps 1 and 2.
2. According to step 3, do multiplication and division. → $6 \cdot 5 = 30$
3. Next, step 4 says to do addition and subtraction. → $2 + 30 = 32$

The answer is 32.

Example: $42 \div 6 \cdot 3 + 4 - 16 \div 2$

$$42 \div 6 \cdot 3 + 4 - 16 \div 2$$
$$7 \cdot 3 + 4 - 16 \div 2$$
$$21 + 4 - 16 \div 2$$
$$21 + 4 - 8$$
$$25 - 8$$
$$17 \checkmark$$

1. Do multiplication and division first (in the order they occur).
2. Do addition and subtraction next (in the order they occur).

Example: $5(2 + 4) + 15 \div (9 - 6)$

$$5(2 + 4) + 15 \div (9 - 6)$$
$$5(6) + 15 \div (3)$$
$$30 + 5$$
$$35 \checkmark$$

1. Do operations inside of parentheses first.
2. Do multiplication and division first (in the order they occur).
3. Do addition and subtraction next (in the order they occur).

Example: $4[3 + 2(7 + 5) - 7]$

$$4[3 + 2(7 + 5) - 7]$$
$$4[3 + 2(12) - 7]$$
$$4[3 + 24 - 7]$$
$$4[27 - 7]$$
$$4[20]$$
$$80 \checkmark$$

1. Brackets are treated as parentheses. Start from the innermost parentheses first.
2. Then work inside the brackets.

Fractions

Changing from one form to another...

Example: **Change the improper fraction, $\frac{5}{2}$, to a mixed number.**

$\frac{5}{2}$ (five halves) means $5 \div 2$.

So, $\frac{5}{2}$ is equal to 2 wholes and 1 half, or $2\frac{1}{2}$.

$$\begin{array}{r} 2 \text{ wholes} \\ 2\overline{)5} \\ \underline{-4} \\ 1 \text{ half} \end{array}$$

Summer Solutions® Problem Solving — Level 7

Help Pages

Solved Examples

Fractions (continued)

Changing from one form to another…

Example: **Change the mixed number, $7\frac{1}{4}$, to an improper fraction.**

1. You're going to make a new fraction. To find the numerator of the new fraction, multiply the whole number by the denominator, and add the numerator
2. Keep the same denominator in your new fraction as you had in the mixed number.

$7\frac{1}{4}$ $7 \times 4 = 28$. $28 + 1 = \mathbf{29}$.

The new numerator is 29.
Keep the same denominator, 4.

The new fraction is $\frac{29}{4}$.

$7\frac{1}{4}$ is equal to $\frac{29}{4}$.

Equivalent Fractions are 2 fractions that are equal to each other. Usually you will be finding a missing numerator or denominator.

Example: Find a fraction that is equivalent to $\frac{4}{5}$ and has a denominator of 35.

$$\frac{4}{5} = \frac{?}{35} \quad \times 7$$

1. Ask yourself, "What did I do to 5 to get 35?" (Multiply by 7.)
2. Whatever you did in the denominator, you also must do in the numerator. $4 \times 7 = 28$. The missing numerator is 28.

So, $\frac{4}{5}$ is equivalent to $\frac{28}{35}$.

Example: Find a fraction that is equivalent to $\frac{4}{5}$ and has a numerator of 24.

$$\frac{4}{5} = \frac{24}{?} \quad \times 6$$

1. Ask yourself, "What did I do to 4 to get 24?" (Multiply by 6.)
2. Whatever you did in the numerator, you also must do in the denominator. $5 \times 6 = 30$. The missing denominator is 30.

So, $\frac{4}{5}$ is equivalent to $\frac{24}{30}$.

When **adding mixed numbers**, follow a similar process as you used with fractions. If the sum is an improper fraction, make sure to simplify it.

Example:

$2\frac{6}{5}$ is improper. $\frac{6}{5}$ can be written as $1\frac{1}{5}$.

So, $2\frac{6}{5}$ is $2 + \frac{5}{2} = 3\frac{1}{5}$.

Help Pages

Solved Examples

Fractions (continued)

When **adding fractions that have different denominators**, you need to change the fractions so they have a common denominator before they can be added.

Finding the **Least Common Denominator (LCD):**
The LCD of the fractions is the same as the Least Common Multiple of the denominators. Sometimes, the LCD will be the product of the denominators.

Example: Find the sum of $\frac{3}{8}$ and $\frac{1}{12}$.

$$\frac{3}{8} = \frac{9}{24}$$
$$+\frac{1}{12} = \frac{2}{24}$$
$$\frac{11}{24}$$

1. First, find the LCM of 8 and 12.
2. The LCM of 8 and 12 is 24. This is also the LCD of these 2 fractions.
3. Find an equivalent fraction for each that has a denominator of 24.
4. When they have a common denominator, fractions can be added.

```
2 | 8, 12
2 | 4, 6      2×2×2×3 = 24
2 | 2, 3
    1, 1
```

The LCM is 24.

To **multiply fractions**, simply multiply the numerators together to get the numerator of the product. Then multiply the denominators together to get the denominator of the product. Make sure your answer is in simplest form.

Examples: Multiply $\frac{3}{5}$ by $\frac{2}{3}$.

$$\frac{3}{5} \times \frac{2}{3} = \frac{6}{15} = \frac{2}{3}$$

1. Multiply the numerators.
2. Multiply the denominators.
3. Simplify your answer.

Multiply $\frac{5}{8}$ by $\frac{4}{5}$.

$$\frac{5}{8} \times \frac{4}{5} = \frac{20}{40} = \frac{1}{2}$$

Sometimes you can use cancelling when multiplying fractions. Let's look at the examples again.

$$\frac{\cancel{3}^1}{5} \times \frac{2}{\cancel{3}_1} = \frac{2}{5}$$

The 3's have a common factor — 3. Divide both of them by 3. Since, $3 \div 3 = 1$, we cross out the 3's and write 1 in each place.
Now, multiply the fractions.
In the numerator, $1 \times 2 = 2$.
In the denominator, $5 \times 1 = 5$.
The answer is $\frac{2}{5}$.

1. Are there any numbers in the numerator and the denominator that have common factors?
2. If so, cross out the numbers, divide both by that factor, and write the quotient.
3. Then, multiply the fractions as described above, using the quotients instead of the original numbers.

$$\frac{\cancel{5}^1}{\cancel{8}_2} \times \frac{\cancel{4}^1}{\cancel{5}_1} = \frac{1}{2}$$

As in the other example, the 5's can be cancelled. But here, the 4 and the 8 also have a common factor: 4. $8 \div 4 = 2$ and $4 \div 4 = 1$. After cancelling both of these, you can multiply the fractions.

REMEMBER: You can cancel up and down or diagonally, but NEVER sideways!

Help Pages

Solved Examples

Fractions (continued)

When **multiplying mixed numbers**, you must first change them into improper fractions.

Examples: Multiply $2\frac{1}{4}$ by $3\frac{1}{9}$.

$2\frac{1}{4} \times 3\frac{1}{9} =$

$\frac{\cancel{9}^1}{\cancel{4}_1} \times \frac{\cancel{28}^7}{\cancel{9}_1} = \frac{7}{1} = 7$

1. Change each mixed number to an improper fraction.
2. Cancel wherever you can.
3. Multiply the fractions.
4. Put your answer in simplest form.

Multiply $3\frac{1}{8}$ by 4.

$3\frac{1}{8} \times 4 =$

$\frac{25}{\cancel{8}_2} \times \frac{\cancel{4}^1}{1} = \frac{25}{2} = 12\frac{1}{2}$

Decimals

When we **round decimals**, we are approximating them. This means we end the decimal at a certain place value and we decide if it's closer to the next higher number (round up) or to the next lower number (keep the same).

Example: Round 0.574 to the <u>tenths</u> place.

There is a 5 in the rounding (tenths) place.   0.574

Since 7 is greater than 5, change the 5 to a 6.   0.574

Drop the digits to the right of the tenths place.   0.6

1. Identify the number in the rounding place.
2. Look at the digit to its right.
3. If the digit is 5 or greater, increase the number in the rounding place by 1. If the digit is less than 4, keep the number in the rounding place the same.
4. Drop all digits to the right of the rounding place.

Example: Round 2.783 to the nearest <u>hundredth</u>.

2.783 — There is an 8 in the rounding place.

2.783 — Since 3 is less than 5, keep the rounding place the same. Drop the digits to the right of the hundredths place.

2.78

Help Pages

Solved Examples

Decimals (continued)

When **multiplying a decimal by a whole number**, the process is similar to multiplying whole numbers.

Examples: Multiply 3.42 by 4.   Find the product of 2.3 and 2.

```
  3.42  — 2 decimal places
x    4  — 0 decimal places
-------
 13.68  — Place decimal point
         so there are 2
         decimal places.
```

1. Line up the numbers on the right.
2. Multiply. Ignore the decimal point.
3. Place the decimal point in the product. (The total number of decimal places in the product must equal the total number of decimal places in the factors.)

```
  2.3  — 1 decimal place
x   2  — 0 decimal places
------
  4.6  — Place decimal
         point so there is 1
         decimal place.
```

The process for **multiplying two decimal numbers** is a lot like what we just did above.

Examples: Multiply 0.4 by 0.6.   Find the product of 2.67 and 0.3.

```
  0.4  — 1 decimal place
x 0.6  — 1 decimal place
-------
 0.24  — Place decimal point so
         there are 2 decimal
         places.
```

```
  2.67  — 2 decimal places
x  0.3  — 1 decimal place
--------
 0.801  — Place decimal point so
          there are 3 decimal
          places.
```

Sometimes it is necessary to add **zeroes in the product** as placeholders in order to have the correct number of decimal places.

Example: Multiply 0.03 by 0.4.

```
  0.03   2 decimal places
x  0.4   1 decimal place
-------
 0.012   Place decimal point so
         there are 3 decimal
         places.
```

We had to <u>add a zero in front of the 12</u> so that we could have 3 decimal places in the product.

The process for **dividing a decimal number by a whole number** is similar to dividing whole numbers.

Examples: Divide 6.4 by 8.   Find the quotient of 20.7 and 3.

```
     0.8
  8) 6.4
    -64
    ----
      0
```

1. Set up the problem for long division.
2. Place the decimal point in the quotient directly above the decimal point in the dividend.
3. Divide. Add zeroes as placeholders if necessary. (examples below)

```
     6.9
  3) 20.7
    -18
    ----
     27
    -27
    ----
      0
```

Examples: Divide 4.5 by 6.   Find the quotient of 3.5 and 4.

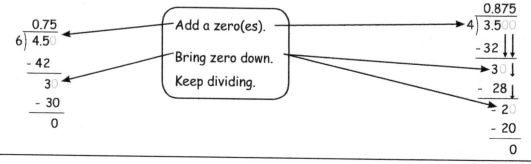

Help Pages

Solved Examples

Decimals (continued)

When dividing decimals the remainder is not always zero. Sometimes, the division continues on and on and the remainder begins to repeat itself. When this happens the quotient is called a **repeating decimal**.

Examples: Divide 2 by 3.   Divide 10 by 11.

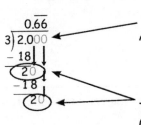

Add zeroes as needed.

This pattern begins to repeat itself (with the same remainder).

To write the final answer, put a bar in the quotient over the digits that repeat.

The process for **dividing a decimal number by a decimal number** is similar to other long division that you have done. The main difference is that we have to move the decimal point in both the dividend and the divisor <u>the same number of places</u> to the right.

Examples: Divide 1.8 by 0.3.   Divide 0.385 by 0.05.

```
     6.
0.3)1.8
    -18
     0
```

1. Change the divisor to a whole number by moving the decimal point as many places to the right as possible.
2. Move the decimal in the dividend the same number of places to the right as you did in the divisor.
3. Put the decimal point in the quotient directly above the decimal point in the dividend.
4. Divide.

```
       7.7
0.05)0.385
     -35
      35
     -35
       0
```

Geometry

The **perimeter** of a polygon is the distance around the outside of the figure. To find the perimeter, add the lengths of the sides of the figure. Be sure to label your answer.

Perimeter = sum of the sides

Example: Find the perimeter of the rectangle below.

```
        9 cm
    ┌─────────┐
5 cm│         │5 cm
    └─────────┘
        9 cm
```

Perimeter = 5 cm + 9 cm + 5 cm + 9 cm

Perimeter = 28 cm

Summer Solutions© Problem Solving   Level 7

Help Pages

Solved Examples

Geometry (continued)

Example: Find the perimeter of the regular pentagon below.

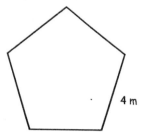

A pentagon has 5 sides. Each of the sides is 4 m long.

P = 4 m + 4 m + 4 m + 4 m + 4 m

P = 5 × 4 m

P = 20 m

Area is the size of a surface. To find the **area of a rectangle or a square**, multiply the length by the width. The area is expressed in square units (ft.², in.², etc.).

area of rectangle = length × width   or   A = L × W

Examples: Find the area of the figures below.

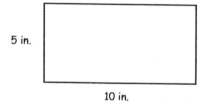

Area = Length × Width

A = 10 in. × 5 in.

A = 50 in.² → 50 square inches

A square has four equal sides, so its length and width are the same.

A = 7 cm × 7 cm

A = 49 cm²

Finding the **area of a parallelogram** is similar to finding the area of any other quadrilateral. The area of the figure is equal to the length of its base multiplied by the height of the figure.

Area of parallelogram = base × height   or   A = b × h

Example: Find the area of the parallelogram below.

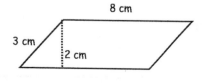

1. Find the length of the base. (8 cm)
2. Find the height. (It is 2 cm. The height is always straight up and down – never slanted.)
3. Multiply to find the area. (16 cm²)

So, A = 8 cm × 2 cm = 16 cm².

Help Pages

Solved Examples

Geometry (continued)

The **circumference of a circle** is the distance around the outside of the circle. Before you can find the circumference of a circle you must know either its radius or its diameter. Also, you must know the value of the constant, *pi* (π). π = 3.14 (rounded to the nearest hundredth).
Once you have this information, the circumference can be found by multiplying the diameter by *pi*.

Circumference = π × diameter   or $C = \pi d$

Examples: Find the circumference of the circles below.

1. Find the length of the diameter. (12 m)
2. Multiply the diameter by π. (12 m × 3.14)
3. The product is the circumference. (37.68 m)

So, C = 12 m × 3.14 = **37.68 m**.

Sometimes the radius of a circle is given instead of the diameter. Remember, the radius of any circle is exactly half of the diameter. If a circle has a radius of 3 feet, its diameter is 6 feet.

Since the radius is 4 mm, the diameter must be 8 mm.
Multiply the diameter by π. (8 mm × 3.14)
The product is the circumference. (25.12 mm)

So, C = 12 m × 3.14 = **25.12 m**.

When finding the **area of a circle**, the length of the radius is squared (multiplied by itself), and then that answer is multiplied by the constant, *pi* (π). π = 3.14 (rounded to the nearest hundredth).

Circumference = π × radius × radius   or $C = \pi r^2$

Examples: Find the area of the circles below.

1. Find the length of the radius. (9 mm)
2. Multiply the radius by itself. (9 mm × 9 mm)
3. Multiply the product by *pi*. (81 mm² × 3.14)
4. The result is the area. (254.34 mm²)

So, A = 9 mm × 9 mm × 3.14 = **254.34 mm²**.

Sometimes the diameter of a circle is given instead of the radius. Remember, the diameter of any circle is exactly twice the radius. If a circle has a diameter of 6 feet, its radius is 3 feet.

Since the diameter is 14 m, the radius must be 7 m.
Square the radius. (7 m × 7 m)
Multiply that result by π. (49 m² × 3.14)
The product is the area. (153.86 m²)

So, A = (7 m)² × 3.14 = **153.86 m²**.

Help Pages

Solved Examples

Geometry (continued)

To find the **area of a triangle**, it is helpful to recognize that any triangle is exactly half of a parallelogram.

The whole figure is a parallelogram.

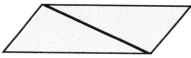

Half of the whole figure is a triangle.

So, the triangle's area is equal to half of the product of the base and the height.

$$\text{Area of triangle} = \frac{1}{2}(\text{base} \times \text{height}) \quad \text{or} \quad A = \frac{1}{2} b \times h$$

Examples: Find the area of the triangles below.

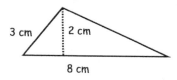

So, $A = 8 \text{ cm} \times 2 \text{ cm} \times \frac{1}{2} = 8 \text{ cm}^2$.

1. Find the length of the base. (8 cm)
2. Find the height. (It is 2 cm. The height is always straight up and down – never slanted.)
3. Multiply them together and divide by 2 to find the area. (8 cm²)

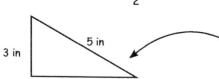

The base of this triangle is 4 inches long. Its height is 3 inches. (Remember the height is always straight up and down!)

So, $A = 4 \text{ in} \times 3 \text{ in} \times \frac{1}{2} = 6 \text{ in}^2$.

Finding the **area of a trapezoid** is a little different than the other quadrilaterals that we have seen. Trapezoids have 2 bases of unequal length. To find the area, first find the average of the lengths of the 2 bases. Then, multiply that average by the height.

$$\text{Area of trapezoid} = \frac{\text{base}_1 + \text{base}_2}{2} \times \text{height} \quad \text{or} \quad A = \left(\frac{b_1 + b_2}{2}\right)h$$

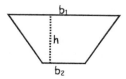

The bases are labeled b_1 and b_2. The height, h, is the distance between the bases.

Example: Find the area of the trapezoid below.

1. Add the lengths of the two bases. (22 cm)
2. Divide the sum by 2. (11 cm)
3. Multiply that result by the height to find the area. (110 cm²)

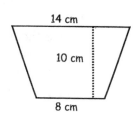

$$\frac{14 \text{ cm} + 8 \text{ cm}}{2} = \frac{22 \text{ cm}}{2} = 11 \text{ cm}$$

$11 \text{ cm} \times 10 \text{ cm} = \mathbf{110 \text{ cm}^2} = \text{Area}$

Help Pages

Solved Examples

Geometry (continued)

To find the **volume** of a solid figure, multiply the area of the base times the height.

Volume of rectangular prism = area of base × height   or   V = l × w × h

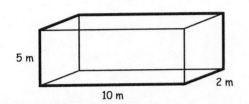

The volume of this rectangular prism is found by multiplying the area of its base (10 m × 2 m = 20 m²) by the height (5 m).

20 m² × 5 m = 100 m³

Ratio and Proportion

A **ratio** is used to compare two numbers. There are three ways to write a ratio comparing 5 and 7.

1. Word form ➡ 5 to 7
2. Fraction form ➡ $\frac{5}{7}$
3. Ratio form ➡ 5 : 7

All are read as "five to seven."

You must make sure that all ratios are written in simplest form. (Just like fractions!!)

A **proportion** is a statement that two ratios are equal to each other. There are two ways to solve a proportion when a number is missing.

1. One way to solve a proportion is already familiar to you. You can use the equivalent fraction method.

2. Another way to solve a proportion is by using cross-products.

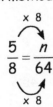

$\frac{5}{8} = \frac{n}{64}$

n = 40

To use Cross-Products:
1. Multiply downward on each diagonal.
2. Make the product of each diagonal equal to each other.
3. Solve for the missing variable.

$\frac{14}{20} = \frac{21}{n}$

$20 \times 21 = 14 \times n$

$420 = 14n$

$\frac{420}{14} = \frac{14n}{14}$

$30 = n$

So, $\frac{5}{8} = \frac{40}{64}$.

So, $\frac{14}{20} = \frac{21}{30}$.

Summer Solutions© Problem Solving — Level 7

Help Pages

Solved Examples

Percent

When changing from a fraction to a percent, a decimal to a percent, or from a percent to either a fraction or a decimal, it is very helpful to use an FDP chart (Fraction, Decimal, Percent).

To change a **fraction to a percent and/or decimal**, first find an equivalent fraction with 100 in the denominator. Once you have found that equivalent fraction, it can easily be written as a decimal. To change that decimal to a percent, move the decimal point 2 places to the right and add a % sign.

Example: Change $\frac{2}{5}$ to a percent and then to a decimal.

1. Find an equivalent fraction with 100 in the denominator.
2. From the equivalent fraction above, the decimal can easily be found. Say the name of the fraction: "forty hundredths." Write this as a decimal: 0.40.
3. To change 0.40 to a percent, move the decimal two places to the right. Add a % sign.

F	D	P
$\frac{2}{5}$		

F	D	P
$\frac{2}{5} = \frac{?}{100}$	0.40	

F	D	P
$\frac{2}{5} = \frac{40}{100}$	0.40	40%

$\frac{2}{5} = \frac{?}{100}$ (× 20)

? = 40

$\frac{2}{5} = \frac{40}{100} = 0.40$

$0.40 = 40\%$

When changing from a **percent to a decimal or a fraction**, the process is similar to the one used above. Begin with the percent. Write it as a fraction with a denominator of 100; reduce this fraction. Return to the percent, move the decimal point 2 places to the left. This is the decimal.

Example: Write 45% as a fraction and then as a decimal.

1. Begin with the percent. (45%) Write a fraction in which the denominator is 100 and the numerator is the "percent." $\frac{45}{100}$

2. This fraction must be reduced. The reduced fraction is $\frac{9}{20}$.

3. Go back to the percent. Move the decimal point two places to the left to change it to a decimal.

F	D	P
$\frac{45(\div 5)}{100(\div 5)} = \frac{9}{20}$		45%

F	D	P
$\frac{9}{20}$		45% = 0.45

F	D	P
$\frac{9}{20}$	0.45	45%

Summer Solutions© Problem Solving Level 7

Help Pages
Solved Examples

Percent

When changing from a **decimal to a percent or a fraction**, again, the process is similar to the one used on the previous page. Begin with the decimal. Move the decimal point 2 places to the right and add a % sign. Return to the decimal. Write it as a fraction and reduce.

Example: Write 0.12 as a percent and then as a fraction.

1. Begin with the decimal. (0.12) Move the decimal point two places to the right to change it to a percent.

2. Go back to the decimal and write it as a fraction. Reduce this fraction.

F	D	P
	0.12	

F	D	P
	0.12 = 12%	12%

F	D	P
$\frac{12(\div 4)}{100(\div 4)} = \frac{3}{25}$	0.12	12%

Often a relationship is described using verbal (English) phrases. In order to work with the relationship, you must first **translate it into an algebraic expression or equation**. In most cases, word clues will be helpful. Some examples of verbal phrases and their corresponding algebraic expressions or equations are written below.

Verbal Phrase	Algebraic Expression
Ten more than a number	$x + 10$
The sum of a number and five	$x + 5$
A number increased by seven	$x + 7$
Six less than a number	$x - 6$
A number decreased by nine	$x - 9$
The difference between a number and four	$x - 4$
The difference between four and a number	$4 - x$
Five times a number	$5x$
Eight times a number, increased by one	$8x + 1$
The product of a number and six is twelve.	$6x = 12$
The quotient of a number and 10	$\frac{x}{10}$
The quotient of a number and two, decreased by five	$\frac{x}{2} - 5$

In most problems, the word "is" tells you to put in an equal sign. When working with fractions and percents, the word "of" generally means multiply. Look at the example below.

  One half <u>of</u> a number <u>is</u> fifteen.

You can think of it as "one half <u>times</u> a number <u>equals</u> fifteen."

When written as an algebraic equation, it is $\frac{1}{2}x = 15$.

Help Pages

Solved Examples

Integers

Integers include the counting numbers, their opposites (negative numbers), and zero.

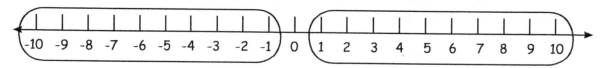

The negative numbers are to the left of zero.   The positive numbers are to the right of zero.

When **ordering integers**, they are being arranged either from least to greatest or from greatest to least. The further a number is to the right, the greater its value. For example, 9 is further to the right than 2, so 9 is greater than 2.

In the same way, -1 is further to the right than -7, so -1 is greater than -7.

Examples: Order these integers from **least to greatest**: -10, 9, -25, 36, 0

Remember, the smallest number will be the one farthest to the left on the number line, -25, then -10, then 0. Next will be 9, and finally 36.

Answer: -25, -10, 0, 9, 36

Put these integers in order from **greatest to least**: -94, -6, -24, -70, -14

Now the greatest value (farthest to the right) will come first and the smallest value (farthest to the left) will come last.

Answer: -6, -14, -24, -70, -94

The rules for performing operations (+, −, ×, ÷) on integers are very important and must be memorized.
The Addition Rules for Integers:

1. When the signs are the same, add the numbers and keep the sign.

$$\begin{array}{r}+33\\++19\\\hline+52\end{array} \qquad \begin{array}{r}-33\\+-19\\\hline-52\end{array}$$

2. When the signs are different, subtract the numbers and use the sign of the larger number.

$$\begin{array}{r}+33\\+-19\\\hline+14\end{array} \qquad \begin{array}{r}-55\\++27\\\hline-28\end{array}$$

The Subtraction Rule for Integers:

Change the sign of the second number and add (follow the Addition Rule for Integers above).

$$\begin{array}{r}+56\\--26\end{array}\ \text{apply rule}\ \begin{array}{r}+56\\++26\\\hline+82\end{array} \qquad \begin{array}{r}+48\\-+23\end{array}\ \text{apply rule}\ \begin{array}{r}+48\\+-23\\\hline+25\end{array}$$

Notice that every subtraction problem becomes an addition problem, using this rule!

Help Pages

Solved Examples

Compound Probability

The **probability of two or more independent events** occurring together can be determined by multiplying the individual probabilities together. The product is called the **compound probability**.

$$\text{Probability of A and B} = (\text{Probability of A}) \times (\text{Probability of B})$$
$$\text{or} \quad P(A \text{ and } B) = P(A) \times P(B)$$

Example: What is the probability of rolling a 6 and then a 2 on two rolls of a die [P(6 and 2)]?

A) First, find the probability of rolling a 6 [P(6)]. Since there are 6 numbers on a die and only one of them is a 6, the probability of getting a 6 is $\frac{1}{6}$.

B) Then find the probability of rolling a 2 [P(2)]. Since there are 6 numbers on a die and only one of them is a 2, the probability of getting a 2 is $\frac{1}{6}$.

So, $P(6 \text{ and } 2) = P(6) \times P(2) = \frac{1}{6} \times \frac{1}{6} = \frac{1}{36}$.

There is a 1 to 36 chance of getting a 6 and then a 2 on two rolls of a die.

Example: What is the probability of getting a 4 and then a number greater than 2 on two spins of this spinner [P(4 and greater than 2)]?

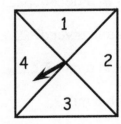

A) First, find the probability of getting a 4 [P(4)]. Since there are 4 numbers on the spinner and only one of them is a 4, the probability of getting a 4 is $\frac{1}{4}$.

B) Then find the probability of getting a number greater than 2 [P(greater than 2)]. Since there are 4 numbers on the spinner and two of them are greater than 2, the probability of getting a 2 is $\frac{2}{4}$.

So, $P(2 \text{ and greater than } 2) = P(2) \times P(\text{greater than } 2) = \frac{1}{4} \times \frac{2}{4} = \frac{2}{16} = \frac{1}{8}$.

There is a 1 to 8 chance of getting a 4 and then a number greater than 2 on two spins of a spinner.

Example: On three flips of a coin, what is the probability of getting heads, tails, heads [P(H,T,H)]?

A) First, find the probability of getting heads [P(H)]. Since there are only 2 sides on a coin and only one of them is heads, the probability of getting heads is $\frac{1}{2}$.

B) Then find the probability of getting tails [P(T)]. Again, there are only 2 sides on a coin and only one of them is tails. The probability of getting tails is also $\frac{1}{2}$.

So, $P(H,T,H) = P(H) \times P(T) \times P(H) = \frac{1}{2} \times \frac{1}{2} \times \frac{1}{2} = \frac{1}{8}$.

There is a 1 to 8 chance of getting heads, tails, and then heads on 3 flips of a coin.

Summer Solutions© Problem Solving   Level 7

Who Knows???

Degrees in a right angle?......................(90)

Degrees in a straight angle?..............(180)

Angle greater than 90°? ....(obtuse)

Angle less than 90°?........................(acute)

Sides in a quadrilateral? ....(4)

Sides in a pentagon? ....(5)

Sides in a hexagon?.................................(6)

Sides in a heptagon? ....(7)

Sides in an octagon?................................(8)

Sides in a nonagon? ....(9)

Sides in a decagon?...............................(10)

Inches in a yard? ....(36)

Yards in a mile?..................................(1,760)

Feet in a mile?....................................(5,280)

Centimeters in a meter?......................(100)

Teaspoons in a tablespoon? ....(3)

Ounces in a pound?...............................(16)

Pounds in a ton?................................(2000)

Cups in a pint?..(2)

Pints in a quart? ....(2)

Quarts in a gallon? ....(4)

Millimeters in a meter? ....(1,000)

Years in a century?...............................(100)

Years in a decade?..................................(10)

Celsius freezing?......................................(0°)

Celsius boiling? ....(100°C)

Fahrenheit freezing?........................(32°F)

Fahrenheit boiling?.........................(212°F)

Number with only 2 factors?..........(prime)

Perimeter? ....(add the sides)

Area of rectangle?............(length x width)

Volume of prism?
..............................(length x width x height)

Area of parallelogram?......(base x height)

Area of triangle? ....($\frac{1}{2}$ base x height)

Area of trapezoid?.....($\frac{base + base}{2}$ x height)

Area of a circle? ....(πr^2)

Circumference of a circle? ....(dπ)

Triangle with no equal sides?........(scalene)

Triangle with all sides equal? (equilateral)

Triangle with 2 sides equal?......(isosceles)

Distance across the middle
of a circle?...................................(diameter)

Half of the diameter? ....(radius)

Figures with the same
size and shape? ....(congruent)

Figures with the same shape,
but different sizes?.........................(similar)

Number occurring most often?........(mode)

Middle number?...............................(median)

Answer in addition?..............................(sum)

Answer in division?........................(quotient)

Answer in subtraction? ....(difference)

Answer in multiplication?............. (product)

Level 7

Problem Solving

Answers to Lessons

Lesson #1

1.
The length is 29 inches.
The width is 5 inches.

$A = l \cdot w \rightarrow 29 \cdot 5 = 145$

w	l = (24 + w)	Product
3	27	81
4	28	112
5	**29**	**145**

2.
The two values are 111 and 333.

111 + 333 = 444
444 ÷ 2 = 222 (average)

3.
The numbers are 25 and 4.

25 + 4 = 29
25 · 4 = 100

4.
The base is 20 cm.
The height is 15 cm.

$A = b \cdot h \rightarrow 20 \cdot 15 = 300$

h	b = (h + 5)	Product
10	15	150
15	**20**	**300**

Lesson #2

1-2.
The average amount the girls brought is $22.

```
  14
  15
  50
  16
+ 15
-----
 110 ÷ 5 = $22
```

The outlier is $50 (Val).

The mean without Val's $50 is $15.

14 + 15 + 16 + 15 = 60
60 ÷ 4 = 15

3.
The length is 32 ft.
The width is 25 ft.

$A = l \cdot w \rightarrow 32 \cdot 25 = 800$

w	l = (w + 7)	Product
20	27	540
22	29	638
25	**32**	**800**

4.
The numbers are 13 and 13.

13 · 13 = 169
13 − 13 = 0

Lesson #3

1	The height is 9 cm. If $bh \div 2 = A$, then $27 \cdot 2 = 54$. $54 \div 6 = 9$	3	The length of the field is 9 ft. The area of the field is 63 sq. ft. If $2l + 2w = 32$, then $32 - 2(7) = 32 - 14 = 18$. $18 \div 2 = 9$ (length) $9 \cdot 7 = 63$ (area)
2	The three percentages are 88%, 92%, and 95%. $92 \cdot 3 = 276$ $88 + 92 + 95 = 275$ (which is very close to 276)	4	The base is 6 mm. If $bh \div 2 = A$, then $33 \cdot 2 = 66$. $66 \div 11 = 6$

Lesson #4

1	213.8°F is the difference in the two temperatures. $134 - (-79.8) = 134 + (+79.8)$ $134 + 79.8 = 213.8$	3	The better scale is 1 in. = 1,000 ft. Rocco's model will be 11 inches high. If he used a scale of 1 in. = 100 ft., his model would be over 9 ft. high.
2	The length of the field is 12 yds. The area is 108 sq. yds. If $2l + 2w = 42$, then $42 - 2(9) = 42 - 18 = 24$. $24 \div 2 = 12$ (length) $12 \cdot 9 = 108$ (area)	4	The three percentages are 65%, 75%, and 95%. $78 \cdot 3 = 234$ $65 + 75 + 95 = 235$ (which is very close to 234)

Lesson #5

1

Fraction	Decimal
1/3	$0.\overline{33}$
2/3	$0.\overline{66}$
3/3	1.00

2

Fraction	Decimal
1/4	0.25
2/4	0.50
3/4	0.75
4/4	1.00

3

Fraction	Decimal
1/5	0.20
2/5	0.40
3/5	0.60
4/5	0.80
5/5	1.00

4

Fraction	Decimal
1/6	$0.1\overline{66}$
2/6	$0.\overline{33}$
3/6	0.50
4/6	$0.\overline{66}$
5/6	$0.8\overline{33}$
6/6	1.00

Lesson #6

1 There are 18 cans of soft drinks, but no mention of how many players are at practice.

2

Fraction	Decimal
1/7	$0.\overline{142857}$
2/7	$0.\overline{285714}$
3/7	$0.\overline{428571}$
4/7	$0.\overline{571428}$
5/7	$0.\overline{714285}$
6/7	$0.\overline{857142}$
7/7	1.00

3 The base is 18 m.
The height is 12 m.

$A = b \cdot h \rightarrow 18 \cdot 12 = 216$

h	b = (h + 6)	Product
10	16	160
11	17	187
12	**18**	**216**

4 The width of the playground is 8.5 ft.
The area is 85 sq. ft.

If $2l + 2w = 37$, then
$37 - 2(10) = 37 - 20 = 17$.

$17 \div 2 = 8.5$ (width)
$8.5 \cdot 10 = 85$ (area)

Lesson #7

1	[diagram showing I-80 running horizontally, with I-376 and I-79 running vertically]	3	**B** Squares always have 4 sides and 4 right angles, and their opposite sides are always equal. All four sides of a rectangle are NOT always equal.
2	A reasonable sum is 4. $\frac{6}{7} \approx 1 \qquad \frac{4}{5} \approx 1$ $\frac{7}{8} \approx 1 \qquad \frac{9}{10} \approx 1$ $1 + 1 + 1 + 1 = 4$	4	Eric's sister is 17. The year after next, she'll be 19. Prime numbers in the teens: 13, 17, 19 17 and 19 are two years apart.

Lesson #8

1	132 is a reasonable product. $65.87 \approx 66$ $2.23 \approx 2$ $66 \cdot 2 = 132$	3	The width of the court is 27 ft. The area is 2,106 sq. ft. If $2l + 2w = 210$, then $210 - 2(78) = 210 - 156 = 54$. $54 \div 2 = 27$ (width) $27 \cdot 78 = 2,106$ (area)
2	The two values are $8.15 and $12.05. $8.15 + 12.05 = 20.20$ $20.20 \div 2 = 10.10$	4	Fraction Decimal 1/8 0.125 2/8 0.25 3/8 0.375 4/8 0.50 5/8 0.625 6/8 0.75 7/8 0.875 8/8 1.00

Lesson #9

1	**260 unique codes can be created.** 26 · 10 = 260 (A-Z = 26 letters) (0-9 = 10 numbers)	3	**22 appointment times** 8:00, 8:20, 8:40, 9:00, 9:20, 9:40, 10:00, 10:20, 10:40, 11:00, 11:20 11 slots · 2 doctors = 22 times
2	**20 combinations** R, W, Y, B, and G are 5 choices for the first draw. Without replacing the first draw, 4 choices remain for the second. 5 · 4 = 20	4	**There are 12 possible combinations.** The choice of one fruit in the a.m. can combine with three different choices from the p.m. 4 · 3 = 12

Lesson #10

1	**There are 10 possible combinations.** N-N-N P-P-P M-M-M N-N-M P-P-N M-M-N N-N-P P-P-M M-M-P N-P-M	3	**16 combinations** Green, Leafy → 4 Root, or Tuberous → 4 4 · 4 = 16
2	Fraction Decimal 1/9 0.$\overline{11}$ 2/9 0.$\overline{22}$ 3/9 0.$\overline{33}$ 4/9 0.$\overline{44}$ 5/9 0.$\overline{55}$ 6/9 0.$\overline{66}$ 7/9 0.$\overline{77}$ 8/9 0.$\overline{88}$ 9/9 1.00	4	Fraction Decimal 1/10 0.10 2/10 0.20 3/10 0.30 4/10 0.40 5/10 0.50 6/10 0.60 7/10 0.70 8/10 0.80 9/10 0.90 10/10 1.00

Lesson #11

1. Aimee bought 4 bags of chips and 4 soft drinks.

It costs $1.25 to buy one of each. (.50 + .75)

5 ÷ 1.25 = 4

2.

Simplest Form	Algebraic Expression
$3n + \frac{1}{2}$	$n + n + n + (8/16)$
$n^3 + 8$	$n \times n \times n + (2^3)$
$2n + 6$	$n + n + (2 \times 3)$

3.

$5.80 is the mean.

```
   3.00
   4.50
  10.00
   5.00
+  6.50
  -----
  29.00 ÷ 5 = 5.80
```

Ben, Allie, Megan, and Kenny have a mean allowance of $4.75.

4. The outlier is $10.00 (Jon).

Lesson #12

1.

Fraction	Decimal
1/11	0.0909
2/11	0.1818
3/11	0.2727
4/11	0.3636
5/11	0.4545
6/11	0.5454
7/11	0.6363
8/11	0.7272
9/11	0.8181
10/11	0.9090
11/11	1.00

2. A reasonable product is 1,696.

211.75 ≈ 212
7.82 ≈ 8
212 · 8 = 1696

3.

Simplest Form	Algebraic Expression
$2n + 6$	$n + n + (18/3)$
$n^2 + 9$	$n \times n + (3^2)$
$3n + 9$	$3(n + 3)$

4. Sabine is 15 yrs old.

15 is a multiple of 3.
14 is a multiple of 7.
16 is the only perfect square in the teens.

Lesson #13

1-2

Terri exercised 185 minutes total.

(Answers are in minutes.)
Mean: **37**   Median: **35**
Mode: **50**   Range: **30**

20 30 35 (50) (50)

```
  M  30
  T  50
  W  20
  Th 35
+ F  50
  -----
     185 ÷ 5 = 37
```

3-4

Cary's brother still owes $8,250.

Month	Balance Owed
	10,000
1	9,750
2	9,500
3	9,250
4	9,000
5	8,750
6	8,500
7	8,250

He will pay off his debt in 33 months.

33 months is less than 3 years.

Lesson #14

1

36 different breakfasts can be created.

4 whole grains, 3 fruits, and 3 milk options

4 · 3 · 3 = 36

2

Will is 24 years old.

Answers will vary.

24 is a multiple of 4 and 6. It is located between a prime number (23) and a perfect square (25).

3

Randy spent 345 total minutes on math.

(Answers are in minutes.)
Mean: **69**   Median: **75**
Mode: **75**   Range: **30**

50 65 (75) (75) 80

4

```
   65
   75
   50
   80
 + 75
  ----
   345 ÷ 5 = 69
```

Lesson #15

1. The cost of the sunscreen was $6.99.

x = cost of the sunscreen

$21.98 = 8 + 2x$
$13.98 = 2x$
$6.99 = x$

2. The cost of the DVD was $12.50.

d = cost of the DVD

$45 = 20 + 2d$
$25 = 2d$
$12.5 = d$

3. C

$7n = 22 - 1$
$7n = 21$
$n = 3$

4. B

12 less → -12
square of ten → 10^2

Lesson #16

1. 676,000 unique license plates

$26 \cdot 26 \cdot 10 \cdot 10 \cdot 10 = 676{,}000$

Answers will vary.
AP439
CW286

2. 10 ways; probability: 10/36

	1	2	3	4	5	6
1	0	1	2	3	4	5
2	1	0	1	2	3	4
3	2	1	0	1	2	3
4	3	2	1	0	1	2
5	4	3	2	1	0	1
6	5	4	3	2	1	0

3.

Simplest Form	Algebraic Expression
$4m^2 + 4m + 24$	$4(m^2 + m + 6)$
$12m + 16$	$4(2m + m + 4)$
$8m + 1$	$4(2m + 1/4)$

4. 8 sheep and 8 chickens.

32 legs + 16 legs = 48 legs

1 chicken and 1 sheep have a total of 6 legs between them.

$48 \div 6 = 8$

Lesson #17

1	7 + -15 = **-8**	3	averages out to: **5 stacks of 4**
2	-10 + 15 = **5**	4	averages out to: **3 stacks of 5**

Lesson #18

1	The cost of the CD was **$15.99.** c = cost of the CD $46.98 = 15 + 2c$ $31.98 = 2c$ $15.99 = c$	3	See table below
2	See table below	4	**9 bicycles and 9 tricycles** 18 tires + 27 tires = 45 tires 1 bicycle and 1 tricycle have a combined total of 5 wheels. $45 \div 5 = 9$

Problem 3 table:

Simplest Form	Algebraic Expression
$7x$	$3x + 2x + 2x$
$5 + 2x$	$3 + 2x + 2$
$3 + 4x$	$3 + 2x + 2x$

Problem 2 table:

Expression	Solution
$2(3 + 4)$	**14**
$2 \times 3 + 4$	**10**
$2^3 + 4$	**12**
$2 \cdot (3 + 4)$	**14**

Lesson #19

1. The length is 113.53 in.

$108.24 \cdot 2 = 216.48$

$443.54 - 216.48 = 227.06$

$227.06 \div 2 = 113.53$ (length)

2. The width is 29 cm.

If $2l + 2w = 112$, then $112 - 2(27) = 112 - 54 = 58$.

$58 \div 2 = 29$ (width)

3. The team gained 8 yards. They did not get a first down.

$5 + -8 = -3$
$-3 + 9 = 6$
$6 + 2 = 8$

4. the 12th day

1 2 ③ ▲4 5 ⬜6 7
▲8 ⑨ 10 11 ⬜▲⑫ 13 14

circle → lily (every 3 days)
triangle → violet (every 4)
square → spider (every 6)

Lesson #20

1. 10:15 am

9:30 - 71°F	9:55 - 76°F
9:35 - 72°F	10:00 - 77°F
9:40 - 73°F	10:05 - 78°F
9:45 - 74°F	10:10 - 79°F
9:50 - 75°F	10:15 - 80°F

2. Titles will vary.

50% Swimming
30% Tennis
10% Biking
10% Golfing

3. The cost of a loaf of bread is $3.50.

b = cost of a loaf of bread

$13 = 6 + 2b$
$ 7 = 2b$
$3.5 = b$

4.

Expression	Solution
$52 + 7 \cdot 4$	80
$5^2 + 7 \times 4$	53
$(5^2 + 7) \times 4$	128
$(5 + 7)^2 + 4$	148
$5(2 + 7) + 4$	49

Summer Solutions© Problem Solving — Level 7

Lesson #21

1	$2\frac{1}{2} \div \frac{1}{4} = \frac{5}{2} \cdot \frac{4}{1} =$ **10**	3	The team gained 11 yards. They got a first down. +8 − 9 = -1 -1 − 7 = -8 -8 + 19 = +11
2	-18°F 32 − 50 = -18	4	Yes The length of each side is 8 units.

Lesson #22

1	The width is 6.9 mm. The area is 64.17 sq. mm. 9.3 · 2 = 18.6 32.4 − 18.6 = 13.8 13.8 ÷ 2 = 6.9 9.3 · 6.9 = 64.17	3	$3\frac{3}{4} \div \frac{1}{8} = \frac{15}{4} \cdot \frac{8}{1} =$ **30**
2	24, 48, and 36 are the 3 values. 24 + 48 + 36 = 108 108 ÷ 3 = 36	4	The game started at 7:30 pm. 9:38 → 9:16 → 8:47 → 8:33 → 7:30 -0:22 -0:29 -0:14 -1:03

Lesson #23

1.

Fraction	Decimal
1/3	0.$\overline{33}$
3/8	0.375
1/9	0.$\overline{11}$
1/6	0.1$\overline{66}$
1/7	0.$\overline{142857}$
3/4	0.75

2. The width is $1\frac{1}{4}$ ft.

5 sq. ft. is a reasonable area.

$4\frac{3}{4} \cdot 2 = \frac{19}{4} \cdot \frac{2}{1} = = 9\frac{1}{2}$

$12 - 9\frac{1}{2} = 2\frac{1}{2}$ $2\frac{1}{2} \div 2 = 1\frac{1}{4}$

3. $2\frac{2}{3} \div \frac{1}{6} = \frac{8}{3} \cdot \frac{6}{1} = 16$

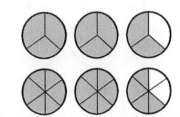

4.

4 stacks of 6

Lesson #24

1. 85°F at noon

60 min. ÷ 10 = 6

79 + 6 = 85

2.

Expression	Solution
48 ÷ 8 − 4 × 1	2
48 ÷ (8 − 4) × 1	12
(48 ÷ 8 − 4) + 1	3
48 ÷ [(8 + 4) × 1]	4

3. $2\frac{1}{4} \div \frac{1}{8} = \frac{9}{4} \cdot \frac{8}{1} = 18$

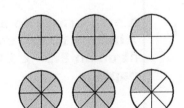

4. The team gained 21 yds. They got a first down.

−9 −9 −9 = −27

−27 + 48 = 21

Lesson #25

1.
The base is 6 cm.
The height is 10 cm.

$bh \div 2 = 30$

$30 \cdot 2 = 60$

$6 \cdot 10 = 60 \qquad 10 - 6 = 4$

2.
Grandma should put the roast in at 3:45 p.m.

$4 \cdot 0{:}20 = 1{:}20$

$1{:}20 + 0{:}15 + 0{:}10 = 1{:}45$

$5{:}30 - 1{:}45 = 3{:}45$

3.
$16 each

$64 \div 4 = 16$

4.
square

All sides congruent means that all sides are the same length, and all angles congruent mean all angles are the same. That makes a quadrilateral with four sides of equal length and four 90° angles.

Lesson #26

1.
1 inch = 40 mi.

A map drawn to a scale of 1 inch = 10 mi. would not fit on an 8.5" x 11" sheet of paper.

2.
(Map showing Chester Ave., Euclid Ave., Carnegie Ave. with E. 55th St. and E. 79th St.)

3.
50 combinations can be ordered.

$5 \cdot 10 = 50$

Answers will vary.
Chicken, Hash Browns, Salad
Tilapia, Applesauce, Salad

4.

Simplest Form	Algebraic Expression
$4x + 3$	$x + 1 + 3x + 2$
$5x + 2$	$x + x + 3x + 2$
$5x + 3$	$2 + 1 + 3x + 2x$

Lesson #27

1. Jo walked 12 mi.

walking:biking = 3:10

$$\frac{3}{10} = \frac{x}{40}$$

$$\frac{3}{10} \cdot 40 = \frac{x}{40} \cdot 40$$

$$12 = x$$

2. **48 is not a perfect square.**

No integer times itself equals 48.

3. 2

$-8 - -10 = -8 + 10 = 2$

4. C

$2(5 + n) = 6 \cdot 3$
$2(5 + n) = 18$
$10 + 2n = 18$
$2n = 8$
$n = 4$

Lesson #28

1. The 2 numbers are 15 and 15.

$15 \cdot 15 = 225$

$15 - 15 = 0$

2. The base is 7 m.

$bh \div 2 = 42$

$42 \cdot 2 = 84$

$84 \div 12 = 7$

3.
Diameter	Circumference
93 ft.	292 ft.
111 ft.	349 ft.
100 ft.	314 ft.

4. cylinder

The net of a cylinder is a rectangle curved around two congruent circles.

Lesson #29

1.

Radius	Circumference
14 cm	88 cm
9 cm	57 cm
11 cm	69 cm

2.

Fraction	Decimal
2/3	$0.\overline{66}$
7/8	0.875
7/9	$0.\overline{77}$
5/6	$0.8\overline{33}$
4/7	$0.\overline{571428}$
2/5	0.40

3.

Simplest Form	Algebraic Expression
$4x^4$	$4x \cdot x \cdot x \cdot x$
$4x + 1$	$x + x + 2x + 1$
$8x + 4$	$4(2x + 1)$

4. 36 different outcomes

$6 \cdot 6 = 36$

Lesson #30

1. Titles will vary.

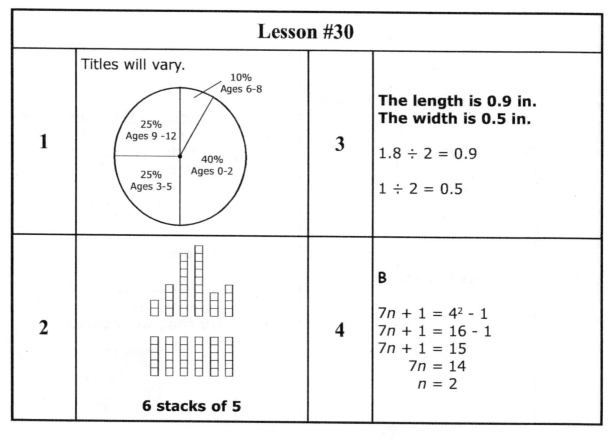

2. 6 stacks of 5

3. The length is 0.9 in.
The width is 0.5 in.

$1.8 \div 2 = 0.9$

$1 \div 2 = 0.5$

4. B

$7n + 1 = 4^2 - 1$
$7n + 1 = 16 - 1$
$7n + 1 = 15$
$\quad 7n = 14$
$\quad\ \ n = 2$